THE ANGRY GENIE

THE ANGRY GENIE

BY KARL Z. MORGAN
AND KEN M. PETERSON

One Man's Walk through the Nuclear Age

UNIVERSITY OF OKLAHOMA PRESS
NORMAN

Library of Congress Cataloging-in-Publication Data

Morgan, Karl Ziegler, 1908–
The angry genie : one man's walk through the nuclear age / by Karl Z. Morgan and Ken M. Peterson.
p. cm.
Includes bibliographical references and index.
ISBN 0-8061-3122-5 (alk. paper)
1. Morgan, Karl Ziegler, 1908– . 2. Medical physics—United States—History—20th century. 3. Manhattan Project (U.S.) 4. Nuclear physicists—United States—Biography. I. Peterson, Ken M. II. Title.
R895.6.U6M67 1998
363.17′995′092—dc21 98-34766
CIP

The paper in this book meets the guidelines for permanence and durability of the Committee on Production Guidelines for Book Longevity of the Council on Library Resources, Inc.∞

1 2 3 4 5 6 7 8 9 10

CONTENTS

CONTENTS

FIGURES

PREFACE

On July 16, 1945, the first atomic bomb is exploded in the desert near Alamogordo, New Mexico. The stem of the explosion rises to great heights before the mushroom flares out. A blinding flash of light, followed by a roar that echoes from the nearby mountains, numbs the observers. The furious energy of the blast derives from plutonium-239 made at Hanford, Washington. For security reasons, only a small number of scientists and reporters are permitted to witness the event.

What they see is unprecedented in human experience. The tower that held the bomb and the sands of the desert beneath melt down like butter. The intense heat turns the sand into small radioactive cinders with a jade-green, glasslike appearance. Iron balls about 1/16-inch in size cover the blast area: they represent the remains of the tower. Later, some bring these radioactive "souvenirs" home.

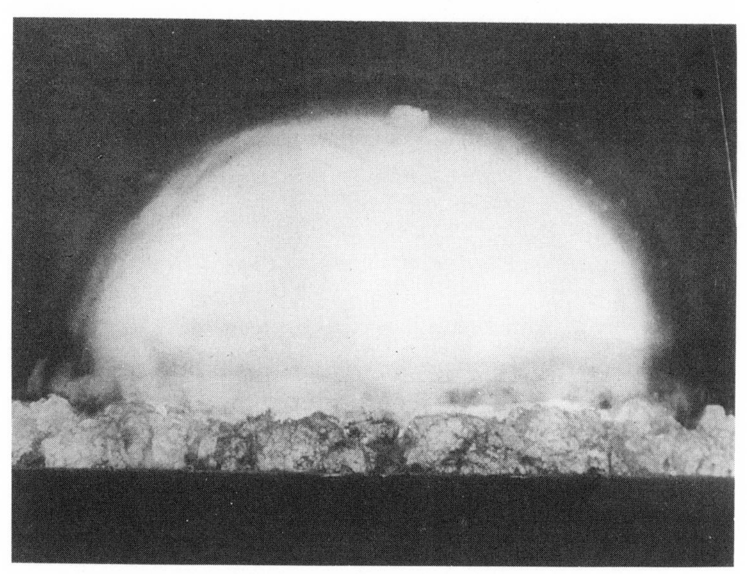

Fig. 1. Trinity Test fireball .053 seconds after detonation. (Photo: Ed Westcott, courtesy of DOE Photography)

Fig. 2. Small radioactive cinders of a jade-green, glasslike appearance, formed by heat generated from plutonium-239 in the Trinity Test. (The individual holding the fused sand is probably not aware that it is still radioactive.) (Photo: Ed Westcott, courtesy of DOE Photography)

This awesome birth of the nuclear age proved far more frightening than I had imagined. Many of us who had worked to make this weapon a success now began to have second thoughts. Was this the beginning of the end or the end of the beginning?

For the next half century I dedicated my life to controlling the most powerful force known to mankind, a cosmic force of the sun and stars now in the fumbling hands of man.

The genie escaped from the bottle that day and can never be put back. For millennia human beings manipulated the energy of molecules (chemical energy) for purposes of good and evil, for life and for death. Now, we are working with the energy of the atom, the same type of energy the sun uses to warm the earth and that fuels the other stars of the heavens.

Hitler and his forces were defeated before the frightening atomic blast at Alamogordo. Atomic power was used—needlessly, I believe, and in revenge—to take the lives of hundreds of thousands of Japanese men, women, and children.

No sooner had World War II ended than the Cold War began. The United States, followed by other nations, rushed to build nuclear power plants. We thought we could put the atom to peaceful use and provide plentiful energy for all people of the earth, but we were foolish. In our mad rush, we neglected safety concerns in pursuit of the overriding objective.

Our primary goal became building so-called breeder reactors, nuclear power plants that would create as much plutonium from plentiful uranium-238 as was consumed. So our plan was to create an endless supply of cheap energy from plutonium—one of the most dangerous substances known.

We gave almost no consideration to the proliferation of nuclear weapons and the danger of plutonium getting into the hands of terrorists. Today hundreds of nuclear power plants exist, producing electricity and weapons materials. None is inherently safe, and weapons-grade plutonium is more accessible to terrorists than ever before.

These reactors are like immense stacks of dynamite in a match factory. They are fraught with limitations that are known to cause accidents. Many of these plants are located too close to large cities. Their "safe" operation requires that at all times they be under the control of well-trained and certified reactor operators. Various mistakes or acts of neglect can be made by reactor operators, which can lead to serious consequences if not immediately corrected. Nuclear power plants are not free from outside events that can cause serious malfunction or accident. They accumulate large amounts of dangerous radioactive particles called radionuclides[1] that remain for years in the reactor—just waiting for an accident to scatter them into the environment. With better design these radionuclides could be constantly removed so they would not accumulate. A safer reactor system, the molten salt thermal breeder, or MSTB, was under early development at Oak Ridge National Laboratory (ORNL) until it was foolishly dropped from support by the Atomic Energy Commission (AEC) and succeeding government funding agencies.

It is very desirable for reactors to have a negative void coefficient in the cooling system, so that if the cooling liquid (usually water) leaks out in an accident, the reactor will shut down. But far too many existing reactors have a positive void coefficient—as did Three Mile Island and Chernobyl.[2]

Some nuclear plants use a moderator, such as graphite, to lower the energy of the neutrons (to thermalize the neutrons). In such reactors large amounts of energy (called Wigner energy) can be accumulated and stored in the graphite waiting to be released suddenly if there is a sharp increase in temperature. This can lead to melting of fuel elements, blockage of the reactor cooling system, and fires in the graphite. It has contributed to several serious reactor accidents. The situation in graphite-moderated reactors is ameliorated by scheduled slow heating of the graphite and proper dissipation of the Wigner energy.

Many of the reactor safely and control devices (relays) suffer from common mode failure. And if individual safety devices can be put out of commission by a problem such as heat, rust, smoke, or water, they may all fail at once during an accident. Most power reactors release thousands of curies of noble gas, tritium (H-3), and hundreds of curies of carbon-14 into the environment each year, thereby causing cancers and genetic defects in unsuspecting members of the general population.

Another disturbing feature of all nuclear power plants is the problem of "brittle fracture." The metals used in reactor components, behind which are millions of curies of radioactive materials, have been chosen and tempered to withstand a wide range of temperatures without becoming brittle and susceptible to fracture from shock. However, when subjected to neutrons, these metals fracture at higher temperatures, which are closer to the operating range. This has resulted in endless repair costs, shutdowns, and high occupational exposure to radiation among reactor operators.

Why did we not spend our time and effort to reduce or eliminate these accident-causing features before we began building power reactors near large human populations?

I am left with a sense of disappointment and anger. The once noble profession of health physics that I helped create over fifty years ago and that was infused with high professional and scientific stature has sunk to a new low.

Today millions of curies of radioactive waste produced by nuclear facilities remain in temporary storage, awaiting development of "safe" long-term storage facilities—while the escaped genie cries, "Stop before it is too late!"

This book is based in large part on the firsthand knowledge of an insider, Karl Z. Morgan, who helped usher in the nuclear age. Coauthor Ken M. Peterson of Wichita, Kansas, is recognized by his peers as one of the best trial lawyers in America. Peterson has a national practice representing plaintiffs in business and personal injury litigation. We have been friends for over ten years. Although Peterson was the primary author of two chapters and served as editor of the other chapters, we agreed to use Morgan's first-person voice throughout.

The result of over eight years of work, this joint effort is intended not only for the general public, but also as a challenge to those who are, or seek to be, health physicists. In spite of the shortcomings and failures of the field of health physics, we are still thankful that it exists as a discipline in the United States and many other countries. The situation in some of the nuclear operations in the former Soviet Union, where raw radioactive waste was discharged directly into rivers and lakes, testifies to the tragic consequences that can occur in the absence of health physics.

ACKNOWLEDGMENTS

We gratefully acknowledge the tireless efforts of Linda Parke, a very gifted individual who has been Ken Peterson's legal secretary for twenty years. Without her, this book would not exist.

Helen Morgan and Jackie Peterson have been particularly understanding of the time denied them in order for us to complete this work. They also served as advisers and proofreaders. The suggestions of Olga Peterson regarding grammar were deeply appreciated.

Other individuals who helped us in our work on this book include Dr. Donald Foster (consultant); Manuel Gillespie (ORNL Photography); Dr. Rufus Richie; Randy Sowell (archivist, Truman Library); Bo Lindell (Swedish scientist); and numerous scientists and technicians who worked under Dr. Morgan while he was director of health physics at Oak Ridge National Laboratory. In addition, we thank the following institutions for assistance: U.K. Atomic Energy Authority; University of Indiana Archives; Franklin D. Roosevelt Library; Wichita State University Library; and Kansas University Library. Dr. Lys Ann Shore's copyediting of the manuscript was of the highest quality.

Finally, we owe a debt of gratitude to our editor, Kim Wiar, and her associates at University of Oklahoma Press for their wholehearted support of our efforts.

KARL Z. MORGAN
KEN M. PETERSON

ABBREVIATIONS

AEC	Atomic Energy Commission
AGR	air-cooled, graphite-moderated reactor
BWR	boiling water reactor
DOE	Department of Energy
EPA	Environmental Protection Agency
ICRP	International Commission on Radiological Protection
ICRU	International Commission on Radiation Units and Measurements
IRPA	International Radiation Protection Association
IXRPC	International X-Ray and Radium Protection Committee
LMFBR	liquid metal fast breeder reactor
MPC	maximum permissible concentration
MSTB	molten salt thermal breeder
NASA	National Aeronautics and Space Administration
NCRP	National Council on Radiation Protection and Measurements
NRC	Nuclear Regulatory Commission
ORNL	Oak Ridge National Laboratory (formerly Clinton Laboratories)
PWR	pressurized water reactor
TMIPHFC	Three Mile Island Public Health Fund Committee
WGR	water-cooled, graphite-moderated reactor

THE ANGRY GENIE

PROLOGUE

My Life before the Nuclear Age

If you know his father and his grandfather, you may trust his son.

MOROCCAN PROVERB

My life as a separate human being began with my birth on September 27, 1907, in the parsonage across the road from the St. Enoch Lutheran Church in Enochville, North Carolina.

Daniel Boone, the most famous of my ancestors, had journeyed with his parents from rural Pennsylvania, about seventy miles west of Philadelphia. Daniel's mother was a Morgan.

My father, Jacob L. Morgan, entered this world on February 7, 1872. While attending North Carolina College, a Lutheran school located in Mount Pleasant, North Carolina, he met and courted my mother, Elizabeth Virginia Clay Shoup. Jacob received his degree in theology and married Elizabeth on May 25, 1903. He dedicated the remainder of his life to the Lutheran ministry.

Reverend Henry Zigler, my mother's grandfather, was one of a long line of Ziglers who had dedicated their lives to the Lutheran ministry. In fact, Martin Luther's mother was a Zigler. Henry Zigler was one of the founders of Susquehanna University, a Lutheran college in Selinsgrove, Pennsylvania.

John Shoup, my maternal grandfather, was also a Lutheran minister. He served during the Civil War as a Union prison guard at Fortress Monroe, near Norfolk, Virginia. He befriended his prisoner Jefferson Davis, president of the Confederacy. When the war ended, Davis gave his good friend his most prized personal possession, a walking cane with a gold handle that had been given to him by a wounded compatriot. It remains one of my cherished possessions.

I was a slow learner in my early school years. One day my father decided he would teach his son how to read. I was scared to death for, as he pointed out words to read, I could not see more than one letter at a time or tell the difference between *will* and *well*. He tried over and over, but finally lost patience. Even several whacks from his belt on my rear end failed to effect the instant cure he desired.

Next to my father, my older sister Gladys possessed the highest innate intelligence of any person I knew growing up. She faced prejudice because of her gender nearly her entire life. Although she ranked first in her premedical school class at Carolina, her professors refused to recommend her to an outstanding medical school like Harvard or Johns Hopkins, simply because she was a woman. Instead, Gladys completed her medical degree at Women's Medical College in Philadelphia. For most of her professional life, she directed a Lutheran hospital in Guntier, India.

On the advice of my parents, I took no science courses while in high school—only engineers needed science, scholars certainly did not. Latin,

Fig. 3. Morgan family photo taken in Salisbury, North Carolina; left to right, Elizabeth Virginia Clay Shoup (mother), Katharine and Lois (sisters), Karl Z. Morgan, Reverend Jacob L. Morgan (father), and Gladys (sister).

Fig. 4. Gladys Morgan, sister of Karl.

French, history, social studies, geography, civics, mathematics: these were the essential courses to become a Lutheran preacher. But even though I had no formal training in science during my school years, my interest in the subject became intense.

Before I reached the age of ten, I was building telephones. At the age of thirteen I made crystal radio sets. While in high school in Salisbury, North Carolina, I constructed the town's first loudspeaker radio. I made my own resistors; my loudspeaker was a coil and vibrating disk mounted on the end of a large seashell, the point of which I sawed off. Neighbors came from miles around to listen.

When I finished high school in 1925, only one institution of higher learning seemed to suit my needs, Lenoir Rhyne College. Situated only sixty miles from Salisbury, it offered what I wanted, a program for pretheological students.

In my freshman year at Lenoir Rhyne I took all the courses helpful to a ministerial student in regular order. I also signed up for physics in the first semester. I became fascinated by this subject, so the next semester I enrolled in modern physics. Over a four-month period I completely changed ambitions. No longer did I want to be a Lutheran preacher, but rather a scientist, preferably a physicist. In my junior year I departed

Lenoir Rhyne to enroll at the University of North Carolina in Chapel Hill, which offered advance degrees in physics and mathematics.

I received my bachelor of arts degree in physics and math from UNC in 1929 and my master's degree in 1930. I went to work for Westinghouse, but soon left in order to enter a graduate program at Duke University.

At Duke, three of us physics graduate students roomed at the same boardinghouse. The distance from our rooms to the physics building on Duke's West Campus was two miles. We could not afford a car, but one of us owned a motorcycle. To make our commute faster and more thrilling, we would tie a long rope to the motorcycle, and while one drove, the other two would rollerskate, hanging onto the rope. We often reached speeds of thirty miles an hour. Occasionally all three of us would ride together on the motorcycle.

When it came time to select a research project for my Ph.D. dissertation, I opted for cosmic radiation. In 1931 this was a comparatively new and unexplored field. Like other graduate students, I labored late into the night pursuing my research, exploring this mysterious radiation from outer space that constantly bombards our planet.

My cosmic ray laboratory consisted of a small wooden shack in the university-owned forest about a mile from the campus. Dr. Walter Nielsen supervised my research. Over a four-year period from 1937 to 1941, he and I, together with the theoretical physicist Lothar Nordheim and various other graduate students, published five articles on cosmic radiation in a major journal, the *Physical Review.*[1] In these papers we presented the results of measurements we had made on cosmic ray scattering and the buildup of secondary radiation as it passed through matter, and information on a new particle of matter called the meson.[2]

It would be impossible to assign proper credit to a single person for the discovery of the fourth basic particle of matter, the meson, which along with the electron, proton, and neutron composes the matter in our universe.[3] The major credit belongs to C. D. Anderson and J. H. Bhabha. Perhaps a small credit should go to our group at Duke and maybe even a smidgen to me individually.

By the time I received my Ph.D. in 1934, the United States was in the depths of the Great Depression and jobs were extremely scarce. Fortunately I received an offer from Lenoir Rhyne College to become chairman of the college's tiny physics department in the fall of 1934.

While teaching one of my classes at Lenoir Rhyne, I took note of one of the women students, a beautiful, little freckled-faced blonde sitting in the front row. Her name was Helen Lee McCoy. I mustered my courage

Fig. 5. Helen Lee McCoy and Karl Z. Morgan on our wedding day, August 2, 1936.

Fig. 6. Karl and Helen Morgan on a 1993 Alaska cruise.

to ask her out, and for our first date we journeyed to a creek near Hickory, North Carolina, where we took off our shoes and waded in the cool water. A lovely lifelong association had begun.

Helen and I were married in 1936. Friends kidded her, telling her she had found an easy way to "get your Ph.D." My father conducted the marriage ceremony on August 2, Helen's birthday, at the pinnacle of Beech Mountain, a site we had selected for its natural grandeur. A few days before the ceremony, I labored to clear the road to the top of the mountain, placing rocks in the mud ruts so that our guests would be able to reach the peak.

The night was clear, and as the sun set, the full moon rose as if to add emphasis to our vow to have and hold each other until death do us part—a vow we have kept through all the years. The next year, on July 29, 1937, our first child was born—a boy whom we named Karl Jr. Karl later acquired a brother, Eric Lee, and two sisters, Joan Elen and Diana.

CHAPTER ONE

The Genie Leaves the Bottle

The Manhattan Project

To be thrown upon one's own resources is to be cast into the very lap of Fortune: For our faculties then undergo a development and display an energy of which they were previously unsusceptible.

BENJAMIN FRANKLIN

HITLER'S CONQUEST OF EUROPE

Starting in 1939 and continuing through 1941, the war in Europe resulted in one loss after another for our allies. On June 3 and 4, 1940, 335,000 soldiers, two-thirds of them British, scrambled onto whatever would float at Dunkirk. They managed to reach what many supposed was only temporary safety on English soil. When France capitulated on June 24, 1940, we expected Hitler's troops to arrive at any moment on the beaches of England. Fortunately he had other plans.

Hitler believed the Third Reich could overrun England at his convenience, especially following smashing blows by the buzz bombs and Luftwaffe. Securing his rear ranks proved more urgent to the Nazi dictator than rushing to deliver the coup de grâce to the British Isles.

On June 22, 1941, Germany launched a surprise attack against Russia. Hitler's diversion provided the United States some feeling of relief and an opportunity to build our defenses. Germany granted us precious time.

During the early war years, while I was teaching at Lenoir Rhyne College, getting my mind off the war seemed impossible. Along with many of my students, I joined the Civil Air Patrol. In addition to my physics classes, I was conducting cosmic ray research at several sites in

North Carolina—in Hickory and Durham, in Linville Caverns near Linville, on Mount Mitchell and Beech Mountain—and also in Colorado, on Mount Evans, near Denver.

My age, marital status, and work as a physics professor made me undesirable as an enlistee and unacceptable as a draftee, so I resigned myself to teaching Civil Air Patrol classes four nights a week. This included courses in navigation, theory of flight, and aircraft engines. Three days a week I took flying lessons in light aircraft and accumulated eighty hours of solo flight. I viewed this work seriously, especially after one of my students drowned when his plane went down off the Atlantic coast where he and others were patrolling for German submarines.

My cosmic ray research in Colorado led to my involvement in what became known as the Manhattan Project. In June 1941 President Franklin Delano Roosevelt established the Office of Scientific Research and Development under the direction of a scientist, Vannevar Bush (fig. 7). Bush was an excellent choice since he was well known within the scientific community that would be called upon to make the project successful. Efforts to develop an atomic bomb were taking place in several different locations, including the University of California at Berkeley, the University of Chicago, and Massachusetts Institute of Technology. When the United States entered the war in December 1941, substantial funding became available and the decision was made to proceed with research on all fronts. Bush decided that the army should be involved in construction activities, and therefore the Army Corps of Engineers opened an office in New York City, naming it the Manhattan Engineer District Office. The Manhattan Project took its name from this office, though it comprised work performed on the uranium and plutonium bombs at several locations around the country.

In the summer of 1941 I worked at Mount Evans in the laboratory of Arthur Holly Compton (fig. 7). Compton, a physicist who specialized in high energy particles and cosmic rays, had received the Nobel Prize in 1927 for his work on the scattering of photons and was one of the best-known U.S. physicists. A square structure 50 feet on a side served as both laboratory and living quarters. The little building was surrounded on all sides by sheets of copper, so that when inside we were effectively in a "Faraday cage." Named after Michael Faraday, an English scientist noted for his work in electricity, this copper enclosure protected us and our Geiger-Müller counters from lightning.

Helen stayed with me part of the time. Some mornings when we exited the laboratory door, she looked like a creature from outer space. Due to the

Fig. 7. Seven months before U.S. entry into the war, this "brain trust" met at Berkeley, California, to discuss the giant cyclotron: left to right, Ernest O. Lawrence, Arthur H. Compton, Vannevar Bush, James B. Conant, Karl T. Compton, and Alfred Loomis. (Photo: Ed Westcott, courtesy of DOE Photography)

electrostatic charge, Helen's long hair would stream up and out in all directions. On these occasions we would promptly return to the shelter of the Faraday cage because of the annoying and stinging electrical discharges from our noses, ears, and eyelashes, and our concern about the danger of lightning. Our fear of lightning proved to be justified when a workman was struck by lightning and killed; this happened on the mountain road only a quarter-mile below our building.

While I was working on Mount Evans, J. C. Sterns, a recognized cosmic-ray physicist and chair of the physics department of the University of Denver, visited me several times. I mentioned to him my desire to move west to the mountains. I told him how at age sixteen I had driven my parents on an 8,000-mile camping trip and had fallen in love with our western states. He immediately expressed excitement, saying that for years he himself had wanted to move east. Sterns suggested we trade positions. He would become head of the physics department of Lenoir Rhyne College, and I would take his place at the University of Denver.

When I returned to Lenoir Rhyne College at the end of the summer, Sterns and I exchanged letters, making preliminary arrangements for the exchange. The prospect of this transition was tremendously exciting—but in the winter of 1942 all correspondence with Sterns suddenly ceased. None of my letters were answered. Disappointed, I concluded that for some unknown reason he had lost interest in his own proposal.

In February 1943, however, I began receiving urgent phone calls and letters from Sterns, Compton, and others, urging me to join them at the University of Chicago. They informed me that they were doing important work relating to my specialty, cosmic radiation. Confident I would be thrilled with the challenge, they urged me to participate in their work, though they told me little about it, emphasizing that they were prohibited from revealing any particulars of this secret project before I actually arrived.

At first I did not show much enthusiasm for what seemed like a pig in a poke. I remained a bit peeved with Sterns because of his long silence. But the presence of Compton, a Nobel laureate, on this secret project aroused my curiosity. Eventually, my imagination began to run wild. Had the group already discovered a new kind of radiation? What could be happening in Chicago that would excite them so much?

Walter Nielsen, chair of the Physics Department at Duke University, and I speculated that the Chicago group was attempting to change matter into energy in conformance with Einstein's theory. But this seemed improbable because of the unlikelihood of attaining the necessary high pressures and temperatures. As I had often explained to my students, "there is enough mass energy in this piece of chalk to supply all the electrical needs for this end of town for many months." Could this energy be used in an atomic bomb of great destructive power? This thought, together with the lure of possible new cosmic-ray discoveries, won me over. A few days later, on April 30, 1943, I boarded the train for Chicago.

INITIAL EFFORTS AT THE UNIVERSITY OF CHICAGO

I settled into a rooming house near the University of Chicago campus on the city's South Side. After completing a preliminary security clearance, I met with Compton, Sterns, Robert S. Stone, and E. O. Wollan. Stone, it turned out, was associate director for health under Compton. Wollan chaired a new program designed to protect lab workers from radiation (fig. 8).

Names like "metallurgical laboratory" and "Clinton Engineering Works" were chosen purposely to be misleading for reasons of security.

Fig. 8. E. O. "Ernie" Wollan (seated), one of the original scientists at the University of Chicago who worked on the Manhattan Project, and Clifford Shull, whom Wollan trained in the diffraction technique. (Photo: ORNL)

As I walked into Compton's office, Sterns greeted me: "Karl, you will be working in the health physics section under Dr. Wollan, who, as you know, is one of us cosmic-ray physicists." He then turned and introduced me to Wollan.

Shocked, I began to make my way toward the door, stating, "There must be some serious mistake. I've never even heard of 'health physics'."

They laughed and chimed, almost in unison, "Hold on, Karl! We had never heard of health physics ourselves, until we invented it a few months ago." They explained that they perceived a serious health problem, which they believed could be handled best by physicists. So this new section at the university was called "health physics."

Stone explained that on December 2, 1942, they had placed in operation a "pile" in a squash court under the abandoned West Stands of Stagg Field on the university campus (fig. 9).[1] The "pile" consisted of a stack of graphite blocks and uranium slugs. It achieved self-sustaining fission at a constant power level and was kept under control with inserted rods containing cadmium. The cadmium captured enough of the neutrons to prevent runaway and an ensuing explosion.

Stone told me that both uranium-235 and a new element of atomic mass 239 and atomic number 94, which we commonly called "Product" or

Fig. 9. The West Stands of Stagg Field at the University of Chicago, beneath which the first nuclear chain reaction took place. (Photo: Ed Westcott, courtesy of DOE Photography)

"Product-239,"[2] have a large fission cross-section, both for thermal and fast neutron fission. Fission literally means a splitting up into parts. The splitting of an atomic nucleus results in the release of enormous energy because in the process a microscopic amount of the mass is converted into energy. For each fission of either uranium-235 or Product, approximately 200 million electron volts of energy is given off. For comparison, the atoms in a red-hot object have average energies corresponding to only 1.6 electron volts.

Either U-235 or Product-239 could be used to make an atomic bomb of tremendous explosive power. We later learned that another way to construct an atomic bomb would be to make U-233 from thorium.

Plans were being made to separate uranium-235 from natural uranium and to produce Product, but my colleagues could not discuss this highly classified program with me in detail until my clearance was complete. I learned later that a new town was to be constructed in Tennessee, called Oak Ridge. And in this new community of scientists, contractors, and military personnel, a large pile called the graphite reactor was to be built to produce gram quantities of Product. Nearby was to be constructed one of the largest single structures in the world at that time, which would produce tons of U-235 by the gaseous diffusion method. I felt I was dreaming to think I might become a part of all this.

Stone reminded me of the tragic history of careless use of radium-226. Painters of radium dials suffered numerous cancers as a result of working in factory shops where, over a period of years, they unknowingly ingested a few millionths of a gram of radium. They had developed the bad practice of pointing their brushes with their lips while using a paint containing a small amount of radium.

The total amount of radium-226 extracted from uranium ores and available for use throughout the world amounted to only about 2 pounds. In a single Product-producing pile, the amount of ionizing radiation would be millions of times that from 2 pounds of radium. Thus, we unknowingly became pioneers in the field of nuclear protection, not just for the radiation worker but for all of humankind.

Constantly aware that the war in Europe was not going well, we realized that whoever developed the atom bomb first would most likely win the war. Unfortunately, we had arrived late for the race.

We had reason to fear that the Third Reich would produce an atomic weapon before we did, since the Germans had already discovered fission. Wollan pointed out that key German scientists, such as Hahn, Strassmann, Meitner, and Frisch,[3] had already published papers indicating successful research on the fission of uranium (fig. 10). Stone stated that German scientists probably had a big lead on us in the race to produce a decisive weapon.

The Chicago group explained to me that Albert Einstein had already expressed to President Roosevelt his concern about Hitler gaining first access to the atomic bomb (figs. 11, 12). They told me that Einstein's letter persuaded FDR to support this secret project (appendix 1).

Both Compton and Stone made clear their total commitment to produce an atomic bomb as soon as possible. They also insisted that the piles be operated safely.

Wollan explained that work on the separation of the radioactive fission products from Product had to be done in shielded enclosures using remote-control procedures. One of our responsibilities as health physicists was to set radiation protection standards for four types of ionizing radiation: alpha, beta, gamma, and neutron.

Alpha particle emitters do not present a problem if kept outside the body, since the alphas can penetrate less than 1 millimeter of soft tissue.

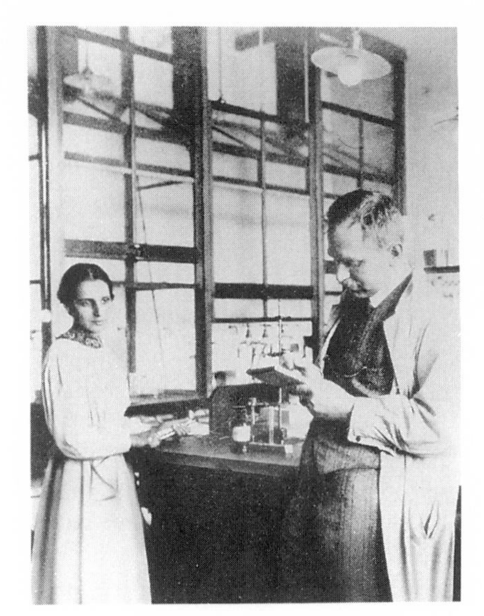

Fig. 10. Nearly two years before I joined the Manhattan Project, German-controlled scientists Lise Meitner and Otto Hahn published scientific papers indicating successful research on the fission of uranium. (Photo: Ed Westcott, courtesy of DOE Photography)

Fig. 11. Albert Einstein, with whom I later spent an evening at his home in Princeton, played a pivotal role in the war effort by persuading FDR to establish what became known at the Manhattan Project. (Photo: Ed Westcott, courtesy of DOE Photography)

Fig. 12. President Franklin Delano Roosevelt, who managed to secretly fund the massive Manhattan Project. (Photo: Ed Westcott, courtesy of DOE Photography)

However, alpha emitters present a serious threat if inhaled or ingested—a real possibility because they attach to dust, food, and clothing. Once alpha-emitting radionuclides enter the body, they cause severe local damage to tissue because they deliver all their energy to relatively few human cells, which may then become cancerous.

Beta particles present similar risks, even though they can penetrate only a little more than 1 centimeter of soft tissue. Gamma or X-ray radiation can penetrate the entire body and alter chromosomes in cells along their tracks.

Fast neutrons penetrate deeply into the body, but their dangers were not fully appreciated until thirty years later, when we discovered that their doses in energy units cause at least thirty times more human damage than gamma rays that deliver the same amount of energy. Neutron radiation was later proposed for use in a small thermonuclear warhead (neutron bomb) that would release large numbers of neutrons intended to kill enemy soldiers without—so it was claimed—destroying buildings and enemy weapons that we could turn against the foe. Later, the U.S. military also seriously considered using waste fission products as an adjunct to chemical warfare.

Our job at Chicago also included developing instruments to survey work areas and monitor worker exposure. Another mandate obligated us to develop safe means to dispose of radioactive waste and to set levels of maximum permissible body burden and acceptable concentrations in air, water, and food for hundreds of new species of radioisotopes. All this was to be done in a way that would prevent radiation injuries, so far as humanly possible. At the same time, however, we understood that the atomic bomb program could not be impeded. It was like being thrown into a cage of lions and instructed not to injure them because they were being trained to destroy the enemy.

From the time I joined the project, I knew that as the first health physicists in the world, our little group of five faced a task of tremendous proportions. Later I realized that the challenge required enormously greater knowledge than any of us originally conceived.

Before the nuclear age, the knowledge of experience with and safe use of ionizing radiation (above about 15 electron volts) was confined to a small group of cosmic-ray physicists and hospital physicists. Worldwide, that was no more than several dozen people.

Cosmic-ray physicists had measured this high-energy background radiation on the earth's surface, on mountaintops, at very high elevations with balloons, in caves deep underground, and at great depths underwater. They had developed photographic emulsions, ion chambers under various gas pressures, cloud chambers, and Geiger counters to measure both primary and secondary radiation. Some of these cosmic radiations are at extremely high energy.

Radiologists and a few hospital physicists in this early period had done observations with photographic films and ion chambers, and related the marginal skin erythema (redness of inflammation) dose to the so-called "threshold" skin erythema dose; after 1940, this unit of measurement was replaced by the roentgen, a unit of air dose.[4] Before 1942, cosmic-ray and medical physicists were about the only persons who used instruments to measure and characterize these high-energy ionizing radiations and who had given some thought to their potential harmful effects. Because of this background, these scientists played a major role in the early development of health physics, a profession designed to protect against the potential hazards of harmful radiations.

Three of us in this first group of health physicists—Herb Parker, Ernie Wollan, and I—were among the few scientists in the world with experience in the development and use of instruments capable of detecting and measuring the various types of high-energy ionizing radiations. As a hospital physicist, Parker had measured the alpha, beta, and gamma radiation levels of radium contamination in hospitals. He had also done some experiments with X-ray film in measuring the radiation exposure of doctors and nurses. Wollan and I brought experience with ion chambers, Geiger counters, and photographic emulsions in our studies of cosmic radiations. Our limited radiation experience included work with alpha, beta, gamma, X-rays, neutrons, and mesons.

Much of our work with Parker involved the development of one of the first photographic film badges, using dental film manufactured by DuPont. We placed filters of lead and cadmium in badges to shield part of the dental film and make blackening of the film by gamma radiation proportional to the roentgen exposure. The amount of blackening of the film was measured with a minometer (photometer). The badge contained an open window that permitted an indication of the radiation dose an individual received from the less penetrating beta radiation.

These early badges enabled us to make a crude estimate of the thermal neutron dose from the neutrons moving slowly enough to be captured by cadmium atoms in the cadmium filter. As each neutron was captured, it produced a gamma ray that, if moving in the right direction, darkened the film behind the cadmium filter.

We were unable to measure fast neutron dose. Some three years later, in 1946, Lyle Borst of the physics division of Clinton Laboratory, later named Oak Ridge National Laboratory, and I developed a thick emulsion film specially designed for insertion into the film badge to measure fast neutron dose. When fast neutrons entered the thick emulsion film, they knocked out protons.[5] The protons, having about the same mass as the neutron, recoiled with half the original energy of the neutron and left a dense ionization track in the film. Later, the proton tracks were counted with a dark-field microscope; the tracks per square millimeter were proportional to the fast neutron dose.

Through trial and error we finally developed pocket condenser meters and pocket fiber electrometers that would not discharge due to rough handling and changing meteorological conditions. These instruments looked like pens and were fastened to shirt pockets. They provided more immediate results of dose received than film badges, since hours of time were saved by avoiding the process of developing film in a darkroom.

Carl Gamertsfelder, another young health physicist, and I spent much of our time at the University of Chicago on the design and calibration of a portable instrument for fast neutron radiation survey. It consisted of two identical cylinders, one containing methane and the other argon gas, both under high pressure. We named it Chang and Eng, for it reminded us of the famous Siamese twins (fig. 13). This was a significant contribution to the

Fig. 13. "Chang and Eng," the portable fast neutron survey instrument designed by Carl Gamertsfelder and me at the University of Chicago. (The instrument got its name because of its two identical cylinders, one of which contained methane and the other argon gas.) (Photo: ORNL)

field of health physics because for the first time it permitted the measurement of fast neutron dose in equivalent roentgen units.[6]

The extent of my ignorance of the true radiation risk I encountered overwhelms me today. While in Chicago, I did not know of any risks of injury from neutron exposure. Often the film in my badge behind the cadmium filter was blackened. During my time there, I received considerably more than a 100 rem dose of neutrons. We know today what I did not know then: that any dose of ionizing radiation causes increased cancers in a given population group, and that on the basis of energy delivered to human tissue, fast neutrons are at least thirty times more harmful than X or gamma radiation (that is, 2 rad of fast neutrons equals 60 rem).

The most important early survey instrument we used was the Lauritsen electroscope. Although awkward to use, this instrument gave us quantitative and reliable readings of milliroentgen (1/1,000 of a roentgen) per hour (mr/hr) of gamma and X-ray dose rate. Sometimes I would support myself on a ladder rung with my left arm, grasping a stopwatch in my left hand, and peer through the microscope of the Lauritsen I was holding gingerly with my right hand. The Lauritsen I used was about a foot high. It was an extremely delicate instrument, one that could not be shaken or dropped. I used a ladder at times to reduce the strong electrical charge from my body while standing on the ground. Such a charge would induce an incorrect reading on the Lauritsen.

Much of my time at the University of Chicago involved efforts to find and develop methods to prevent radiation exposure and to determine what would be a "safe" level of exposure. Little did I suspect that there is *no* "safe" level of exposure to radiation. I also attempted to estimate the types and extent of chronic injury that might be expected from a given exposure.

We gave little consideration to internal dose from body intake of radionuclides because almost nothing had been published on the subject except some information on high-level exposure to radium. We focused primarily on preventing the radiation syndrome (acute death) from very high exposures. Our secondary concern was to prevent acute external radiation damage, such as skin erythema. Unfortunately, we accepted the threshold hypothesis: that so long as we avoided the skin-reddening threshold dose, all of us were safe. We erroneously thought that the system of macrophages found in human bone marrow, liver, and spleen would repair any injury within a few days. Radiation-induced cancer, lens cataracts, life shortening, and genetic damage never occurred to us as a possibility except at very high doses.

We worked seven days a week and often around the clock in our rush to help defeat Hitler. It was a hot Chicago summer in 1943—without air conditioning—and I spent the first two months also without my family. When Helen and our three children joined me, we moved into a furnished apartment. Life was much better when we were all together, but we still didn't care for city life and longed for the war to be over so we could return to the South.

A NEW SCIENTIFIC COMMUNITY—OAK RIDGE

One early September day in 1943, Martin D. Whitaker visited me at my Chicago laboratory. Thirteen years earlier we had attended graduate school together at the University of North Carolina. Whitaker said, "Karl, I have been asked to become director of a national nuclear laboratory that is being secretly set up on some farmland near Clinton, Tennessee. Would you care to move south and work with us?" Overjoyed, I exclaimed, "Yes, yes." Now I would be able to get away from this big city and settle closer to home.

Later that month I drove my family back to our home in Hickory, North Carolina. A few days later I boarded the train for Knoxville. The last twenty miles I traveled by bus past cornfields and pastures to a site six miles west of Clinton, Tennessee. I had no idea that within a few months, eighty thousand people would reside near the bend of the river ahead of me.

As soon as the bus crossed the Clinch River over Solway Bridge, I observed a small concrete structure with narrow, vertical windows, obviously intended to accommodate machine guns. As we approached, men in military uniform emerged and stopped us for a search. Each passenger was required to fill out forms and provide detailed identification before reboarding the bus. Then we proceeded into a bus staging area. Everywhere I looked I saw GIs, bulldozers, trucks, stacks of building materials, and newly constructed wooden barracks.

After an initial interrogation, they assigned each of us to a dormitory and instructed us how to get to work the next morning. They emphasized the absolute necessity for tight security. Only a few top officials, such as Compton, Whitaker, and Stone, drove company cars. The rest of us walked, except for the bus rides to and from our assigned workplace.

A large wooden structure about 150 feet by 100 feet with no room partitions served as my dormitory. Somehow they managed to squeeze a hundred cots into this enclosure, with narrow alleys between each row. A year later this structure was converted into the Alexander Hotel.

One chair and a small cabinet stood beside each cot. Before the first week ended, most of us had at least three pairs of shoes under our cots. Each morning the mud on one pair was dry enough to crack off when we would pound our shoes on the back steps of the building.

Two small toilets and a large common shower room have to meet the needs of a hundred men. The workers often pushed and cursed one another during this hectic dressing procedure. Men waited in long lines at each toilet. Others lined up to shower or to wait for the mirror so they could shave.

We ate breakfast at a nearby army-style cafeteria. Then we walked to the designated area to board buses marked "Construction Area," "Power Plant," "Y-12," "K-25," "X-10," and so forth. I would board the X-10 bus, a typical army bus with hard seats and small windows, appropriately called a cattle car. We would travel ten miles on a rough dirt road to reach X-10, one of the three Oak Ridge locations under construction (the others were Y-12 and K-25; see figs. 14, 15, 16).

During my first month of the military lifestyle in this little town, Helen and our three young children stayed at home in Hickory, packing up our belongings for the move to Tennessee. Meanwhile, our assigned house was under construction at Oak Ridge. As soon as I received word that our

Fig. 14. Wartime aerial view of X-10 under construction. (Photo: ORNL)

Fig. 15. Wartime view of groundbreaking for the Y-12 operation. (Photo: ORNL)

Fig. 16. Wartime view of the K-25 operation. (Photo: ORNL)

house was finished, Helen started out for Knoxville with our youngsters. That afternoon, I was told that not enough painters could be found to paint our house as well as the hundreds of other new homes being constructed. I hurriedly took the bus to Knoxville, waited along the side of the road, and was able to flag down Helen so I could accompany my family to the assigned hotel.

For two long weeks Helen remained alone with the children at the hotel. Excitement filled me when construction of our home was finally completed—now my family would be sleeping under the same roof again. Before I knew it, trucks arrived from Hickory with our furniture and other belongings.

Families of scientists, engineers, administrators, and others were also settling into their new homes. The population of our community surged to eighty thousand people, most of them carpenters, roofers, plumbers, government officials, electricians, truck drivers, and technicians.

The vast majority of houses in this strange new Tennessee town were built on a mountain ridge covered with oak trees, which gave the community its name, Oak Ridge.

Dwellings were constructed from cemestos, a composition of cement, asbestos and sand with wood fiber insulation sandwiched between. There were six floor plans, labeled A through F in order of increasing size and desirability. Only the most important individuals—such as Whitaker, Stone, Compton, and Eugene Wigner—lived in F houses. Wollan, Parker, and I received D house assignments. Eight years later, as director of the Health Physics division, I graduated to the top of the pecking order with an F house.

Construction workers lived in barracks in the valley below the town. I was embarrassed that our government relegated blacks to less desirable homes in a lower part of the valley along Poplar Creek, known as the Scarborough area.

After killing my third copperhead snake in the front yard of our D house, I never again questioned our pioneer status. One was crawling toward my barefoot, three-year-old son, Eric, when I dispatched it with an always-handy garden hoe. Following Helen's recommendation to the town manager, boardwalks were built through the woods where the cemestos homes were located to the schools and down to the one church building, stores, and staging areas in the valley.

In the early period the town business section, known as Jackson Square, had only one grocery store, a drugstore, a post office, and a medical facility. Long lines waited at each. Helen would sometimes stand in line to purchase a pound of butter, only to find it sold out. We could not afford to be choosy in the stores.

A gasoline stamp and twenty-five cents would buy a gallon of gasoline, but the stamps were in short supply. We learned to walk wherever we wanted to go. Eventually, a clothing store opened in town, but no one purchased fancy clothes. Even laboratory scientists seldom wore a coat and tie. In the winter we wore sweaters to keep warm. Initially, all the house furnaces operated on soft coal, throwing the entire town under a pall of smoke. The U.S. government owned all the homes and businesses, as well as the land on which they stood.

The town grew so fast, and the population was so young, that there were not enough kindergartens and elementary schools. All the schools were racially segregated.

The army hospital—especially the maternity wing—was always filled to capacity. New babies arrived—nearly a dozen every day. Some local residents claimed this resulted from the loud noise of bulldozers and hammers that would wake up couples at daybreak each morning. The baby boom lasted for the first ten years.

A favorite gripe among Oak Ridgers was the incompetence of the government and its major contractor, Roane Anderson. The company got its name because the security area followed a bend of the Clinch River and covered a portion of Roane and Anderson Counties. For the most part, however, both the government and Roane Anderson performed well in housing and accommodating the early residents.

Three principal operations have continued at Oak Ridge for decades: (1) X-10, the laboratory facility, variously called Clinton Laboratories, Oak Ridge Lab, and Oak Ridge National Laboratory (ORNL); (2) Y-12, for a few years the plant for electromagnetic separation of uranium-235 from natural uranium, and later a nuclear weapons manufacturing and research facility; and (3) K-25, still one of the largest structures in the United States, used for the separation of uranium-235 from natural uranium by the method of gaseous diffusion.

Both safety and security required that the three principal operations be separated. For example, if the graphite pile at X-10 were to blow up, it would be important that it not seriously damage Y-12 or the uranium-235 separation operations at K-25. Therefore, the plants were built about eight to ten miles apart. Workers at any one of these three plants were not to know what was going on at the other two. Inevitably, the scientists found out. For example, all the transformers and other equipment supplying a tremendous amount of electricity to Y-12 were a complete giveaway of what was happening there. The cooling tower stacks for the X-10 graphite reactor, which never vented anything but hot air, must have caused even the junior physicists to realize something extraordinary was happening.

Parker directed the initial health physics program at X-10. Those from the Chicago group who joined the Clinton Laboratories health physics program were H. M. Parker, C. C. Gamertsfelder, J. C. Hart, and myself.[7] E. O. Wollan transferred from Health Physics to the Physics Division of the lab when he arrived.[8] In addition, twelve DuPont trainees, fifteen GIs, and ten women technicians were added to our initial health physics group. Nearly all the women technicians and secretaries lived in the small towns surrounding Oak Ridge. A few lived as far away as Knoxville (thirty miles).

The University of Chicago supervised X-10 until July 1, 1945. Shortly afterward Monsanto Chemical Company took over and ran X-10 and K-25 until March 1948. Later Union Carbide and still later Martin Marietta in turn took over operations. The Tennessee Eastman Company operated Y-12 until May 1, 1947. But after producing a quantity of uranium-235 by the electromagnetic separation technique that would have been sufficient to supply the explosive material for the bomb used at Hiroshima, its operation ceased, and Tennessee Eastman left Oak Ridge. Y-12 was then converted into a producer and tester of other components for nuclear weapons.

K-25 continued operations as a gaseous diffusion plant until 1979, when this U-235 enrichment operation was shut down. Since 1979 K-25 has been used mainly in cleanup and waste disposal activities (such as the recovery of hundreds of pounds of mercury that had been carelessly discharged into Poplar Creek). As of 1998, X-10 and Y-12 both remain in operation, X-10 as a national laboratory and Y-12 as a weapons research and atomic bomb disassembly facility. The current operations at Y-12 include making fissile material, such as U-235, impotent for safe storage purposes and to meet nonproliferation requirements.

Standard nuclear weapons use either U-235 or Pu-239 as their principal explosive fissile material.[9] These materials also provide energy for nuclear power plants. While power plants using U-235 have concentrations of approximately 5 percent uranium, weapons-grade uranium requires a concentration of 93 percent or higher. Today the Y-12 plant makes weapons-grade uranium-235 impotent by metallurgically blending it with depleted uranium. Ironically, having spent billions of dollars to concentrate this U-235, we are now spending millions more to dilute it with U-238.

Little can be done to make Pu-239 impotent. Weapons-grade plutonium can be mixed with depleted uranium and used in costly, specially designed reactors. But it would still be possible for terrorists to separate the plutonium, thus making it weapons-grade once again.

Much of the operations and gaseous diffusion work at K-25 remains in limbo because plenty of U-235 has been stockpiled, and more comes from disassembly of nuclear warheads. Once a precious commodity, now U-235 is a liability.

During the war years, we health physicists spent little time in our headquarters. Instead we were on site in the graphite pile building and chemical separations facilities where we monitored radiation exposures.

Although Dr. Whitaker directed the X-10 Clinton Laboratories operation, executives of the Monsanto Company, such as Mr. Lum, would sometimes pretend they were the bosses. Similarly, Colonel Mike Murphy seemed to believe he was in charge, probably because he outranked the hundred or so GIs assigned to the laboratory. Titles meant little to us. We clearly understood our principal task, and we proceeded at full steam knowing our German counterparts were diligently pursuing the same goal. We knew that as early as September 1939 Germany had established a special office for the military application of nuclear fission. The Germans' chain reaction experiments were carried out with measurements using carbon as a neutron moderator. Fortunately for us, these measurements later proved erroneous.[10] The Germans used commercial carbon, which contained enough iron-cobalt to skew the measurements. These incorrect results led physicist Werner Heisenberg to recommend the use of heavy water (^2H_2O) as a moderator.[11] German dependence on this extremely scarce commodity was a primary reason why their experiments were never successfully concluded.

RESEARCH AT ORNL: OUR MISTAKEN BELIEF IN A THRESHOLD HYPOTHESIS

Initially, I considered that I had three principal self-imposed assignments: first, to measure the fast neutrons at various points about the pile; second, to look for new kinds or unpredicted types of dangerous radiation; and third, to educate and train the scores of GIs, DuPont trainees, and other new employees in matters of radiation protection, instrumentation, radiobiology, and radiation physics. Later, some of these trainees went on to direct health physics departments at other sites. For example, Jack Healy led the health physics operations at Los Alamos National Laboratories, and C. Patterson at the Savannah River Plant.

With Wollan's encouragement I built a large set of my cosmic ray equipment, which was set up at one face of the Oak Ridge pile ready to take measurements soon after it began operating.

I placed 2 cm of lead in front of four Geiger counters arranged out of alignment so that a single high-energy particle could not penetrate all four counters. I hoped to get no coincident counts, because that would indicate scattering and showers produced by mesons or some other unpredicted radiation that could be very harmful to workers in the pile building. To my chagrin, as the power of the reactor increased, I began to get a few coincident counts.

I discussed my findings with Wollan, who by now had moved to the Physics Division. We reasoned that the flux (frequency of production) of these particles behaving like mesons was so small that at planned pile operations it probably would not deliver a serious radiation dose to workers in the pile building.

We planned to send a letter to the editor of the *Physical Review* to describe our findings. First, however, I continued collecting more data. The coincident counts now seemed erratic and even continued for a short while after the pile was shut down. I made measurements with my portable survey meters and also measured air samples. I found that the air in the pile building was permeated with argon-41 gas, a radionuclide with a half-life of 1.82 hours that gives off a 1.29 MeV gamma ray. This was setting off my individual counter so fast that the counters occasionally gave a coincident count. Today, with resolution times of 10^{-10} seconds, this would not have been a problem, but the equipment available then had resolution times of only about 10^{-2} seconds, so that counts within 1/100 of a second appeared coincident. I was thankful I had not rushed to publish something I would later have had to recant.

My second assignment required me to measure the fast neutron flux escaping through the 6 feet of concrete shielding at several critical locations around the pile. After spending two long days working with a large Chang and Eng, and all its associated equipment, I was ready to take my first data.

That same day the Nobel laureate Enrico Fermi unexpectedly visited my work site (fig. 17). He stopped and asked what I was working on. I replied, "I'm measuring the fast neutron flux coming through the pile shielding and have just gotten my first results."

Fermi said, "Don't tell me what you got, let me see what it should be at this location." He pulled out a pencil from his coat pocket and scribbled on the back of an envelope for a few minutes. Then he announced, "Karl, it should be about two fast neutrons per square centimeter per second at this location." And in fact, that was exactly what I had measured.

I asked myself how anyone could be that brilliant. How could he know how much hydrogen and other elements were present in the 6 feet of concrete at this location?[12]

Fermi possessed a prodigious memory. Not only had he memorized long portions of Dante's *Divine Comedy*, he would also startle coworkers, such as Albert Wattenberg, with his mastery of physics. Wattenberg, who helped Fermi create the first self-sustaining pile, commented, "I probably measured the cross-sections for about 70 different elements [to discover their ability to absorb neutrons]. In the course of discussions when something came up, Fermi remembered, off the top of his head, all the values of those cross-sections."[13]

Such gifted nuclear physicists of that day as Fermi, Szilard, Wigner, Wheeler, Einstein, Compton, Stone, Gofman, Lawrence, Oppenheimer, Urey, Zinn, Weinberg, and others, provided the United States with more scientific capacity than any other nation in history. Also important was our close cooperation with leading scientists in the United Kingdom and Canada.

Fig. 17. Enrico Fermi, one of the most brilliant scientists I worked with, who played a major role in directing the creation of the world's first atomic "pile" under the West Stands of Stagg Field at the University of Chicago. (Photo: Ed Westcott, courtesy of DOE Photography)

As part of my third self-imposed assignment—training health physicists and educating laboratory employees regarding the risk of radiation exposure—I gave talks before assemblies of employees, led individual discussions, and distributed written material. Admiral Hyman Rickover and his officers attended a few of my classes. Later, the admiral became the founder of our nuclear navy. Many other leaders in what became the nuclear industry attended my lectures.

Early on we all accepted the threshold hypothesis. If one had a little skin reddening, there was no need for concern. But how much was too much?

Skin erythema from beta radiation represented a special early concern at X-10 because some employees refused to wear heavy leaded gloves and even handled uranium slugs with their bare hands when we weren't looking.

Parker and I independently calculated the surface beta dose from a uranium slug to be about 270 milliroentgen per hour. We conducted what we believed to be a safe human study in 1944 to check our calculations and, we hoped, confirm the safety of some of the laboratory and engineering operations. If we had known then what we know today, this study would never have been undertaken.

We obtained some circular plaques of sulfur about 1 inch in diameter and 1/4-inch thick, and placed them in the X-10 pile for a few weeks.[14] Then we taped these plaques to the arms of nurses, doctors, and members of our health physics group, and left them there for a week or more. After a week we noted only slight reddening, so we kept them on our arms for a period of ten days. By then distinct red marks appeared on our arms. We worried when some of the nurses showed a few blisters and slight ulceration at the exposure sites. Now we knew that this erythema was trickier than a sunburn, since it originated deep below the skin surface and, unlike sunburn, developed slowly over a period of a week or more.

Although we issued a film badge to everyone exposed to ionizing radiation, some employees left their badges at home or on the desk in their office. After a week of arguing with persons in charge of security, we arranged to combine the film badge with the security badge. Now the film badges contained space for a picture of the employee and his or her security number. No one dared be seen in the restricted area without the proper badge.

Ironically, some scientists, especially physicists and chemists, displayed carelessness and at times even scorn for the issue of radiation protection. They thought we were exaggerating the risk of radiation exposure. College professors are unaccustomed to being told not to do something by a "technician." A few months after startup of the pile, Whitaker, the laboratory director, received some distinguished visitors. He showed them the pile where some interesting experiments were taking place.

On one side of the pile a large opening ran through the 6 feet of concrete shielding. The opening contained a large tank normally filled with water in which boron was dissolved to absorb neutrons and serve as shielding against both gamma and neutron radiation.

Due to some special experiments that were under way, the tank was empty. As Whitaker and his guests entered the pile building, they ignored the Health Physics Department's yellow ribbons bearing "Danger" warnings and "No Admittance" signs placed conspicuously in the area near the empty water tank. They went on to tour the reactor area. Our director thereby exposed himself and his guests to a life-threatening direct beam from the operating pile. Horrified to see them in the prohibited area, we rushed them out. We were never able to determine the dose they received because Whitaker and his guests had also ignored laboratory rules by not wearing film badges.

Fortunately, they spent only seconds in this very dangerous zone. Had they tarried there much longer, they surely would have received a fatal dose. We estimated that each received a dose equivalent to approximately 50 roentgens.

In some respects this accident proved fortunate, since thereafter the laboratory director strongly supported all health physics rules. Everyone who might conceivably enter a radiation zone now wore a film badge and risked being fired if he or she ignored the yellow ribbons and warning signs.

Early in 1944 our laboratory prepared to ship large curie amounts of barium-140 to the Los Alamos site in New Mexico for use in some of their weapons studies. Barium-140 is primarily a beta emitter that is easily shielded, but its daughter, lanthanum-140, is a high-energy gamma ray emitter. Barium-140 has a half-life of 12.75 days, so during truck shipment to Los Alamos and after arrival it became an intense gamma radiation source used for experimental purposes. Some of the truck drivers unknowingly received large doses of ionizing radiation.

One day just outside my office at X-10, I heard someone yelling and ran out to investigate. Colonel Murphy was "dressing down" one of my GI employees because the GI had criticized him for cleaning out the burning embers of his tobacco pipe on the running board of the truck—which was loaded with 400 curies of barium-140, carefully shielded, for shipment to Los Alamos. This constituted by far the largest accumulation of curies of radioactive material in the world.

I, too, gave the colonel a piece of my mind, informing him that had the truck caught fire, it could well have spread high levels of radioactive contamination over not just the entire laboratory but a large portion of the entire Oak Ridge area. Such a senseless act could have put thousands of residents at risk of serious injury or death. This ruffled his feathers so much that he brought the matter to Whitaker's attention. It was a shock to him when Whitaker commended me for my action!

A NEW RESPONSIBILITY: WARTIME DIRECTOR OF HEALTH PHYSICS

Nine months after our group from Chicago arrived in Tennessee, Parker, Gamertsfelder, and Hart moved to Richland, Washington, where Parker directed the health physics program at Hanford.

At this time I became director of the Health Physics Division at ORNL, charged with the responsibility of containing and controlling the various types of radiation created by our efforts in the development of the first atomic bomb.

In order to have a reference point in future years, we measured the natural background of ionizing radiation at the laboratory and in the Oak Ridge area before the operation of our reactor or pile. The average background level was about 0.008 milliroentgen per hour and served as our baseline in following years. Even when the background level increased by factors of 10 to 20, we were not concerned because of our belief in the threshold hypothesis. The city of Oak Ridge registered up to 60 times the natural background level for a few days on several occasions.

Initially referred to as the Clinton Pile or Oak Ridge Pile, the reactor eventually became known as the Oak Ridge Graphite Reactor. Technically, it was an air-cooled, graphite-moderated reactor (AGR). Today it is a prized historical relic. It consists of a large stack of graphite blocks about 30 feet on a side and 30 feet high, contained inside a 6-foot-thick rectangular shell of concrete (fig. 18). Far larger than the first pile built in December 1942 under the West Stands of Stagg Field at the University of Chicago, it served as a pilot plant for demonstrating the feasibility of building and operating the large plutonium-producing, water-cooled, graphite-moderated reactors (WGRs) at Hanford, Washington, and the AGR at Windscale, England.

Our AGR commenced operations or became "critical" on November 4, 1943, a little over a month after the the arrival of the Chicago health physics group. (A pile becomes critical when a nuclear chain reaction becomes self-sustaining.) It achieved full power in stages and eventually, in May 1944, reached the thermal power level of 1,800 kilowatts. Later it operated at somewhat higher power levels. By March 1944 several grams of deadly plutonium had been recovered from the reactor and associated chemical operations. The development of various methods of removing this plutonium from the aluminum, uranium, and fission products proved very helpful in giving operations at Hanford a head start.

For the most part the graphite reactor operated without major incidents. One exception was the case of Whitaker and his guests, mentioned

Fig. 18. The Oak Ridge Graphite Reactor, initially known as the Clinton Pile. (Photo: ORNL)

earlier; another occurred because of a reckless act of another scientist, Lewis Slotin.

The graphite reactor had a large opening on top, and in this sat a water tank about 6 x 6 x 6 feet. During normal operations, the tank contained borated water, which provided adequate shielding of persons who might be working on top of the reactor for a short time. Slotin set up an experiment at the bottom of this water tank, where it was exposed directly to very high levels of radiation.

One Friday afternoon, his equipment malfunctioned. Slotin asked that the reactor be shut down so that he could drain the tank and repair his equipment. But several other researchers also had experiments under way, which would be ruined if the power level was changed ahead of schedule. We asked Slotin to wait until Saturday afternoon, when the reactor was scheduled to be shut down for a fuel change. But during the night Slotin stripped down to his shorts, dove to the bottom of the water tank, and repaired his equipment. He did not wear his film badge, so we could only estimate his radiation dose to be at least 100 roentgens. I was appalled by his recklessness.

Some months later Slotin transferred to the Los Alamos site. On May 21, 1946, while working in the Los Alamos Atomic Bomb Laboratory, Slotin arranged a stack of uranium-235 blocks in such a way that he could vary the snugness of the blocks very slightly, using a screwdriver, and listen to a Geiger counter nearby as it clicked faster and faster, and then went wild when the blocks were slowly allowed to come to a tighter assembly. This was one way to "twist the dragon's tail."

Unfortunately, on this occasion, the screwdriver slipped, and the blocks fell tightly together, creating a "critical assembly." A critical assembly occurs when a sufficient mass of fissionable material, such as Pu-239 or U-235, is brought together or compacted to create an explosive chain reaction. When the assembly is tight enough, neutrons find it difficult to escape and thus build up. The explosion occurs because an enormous amount of energy is brought together in an extremely short period of time, heating the assembly and the surrounding air, and the heated air expands abruptly.

In a flash Slotin received a neutron dose of 287 rad and an unknown amount of gamma exposure (he was not wearing his film badge at the time of the accident). He died nine days later, at the age of thirty-five, having suffered excruciating pain. Several other people received large exposures, but survived because Slotin's body shielded them as he delayed his departure to unscramble the assembly and offer them some protection. Slotin became the first person in history, but not the last, to die from radiation exposure at a nuclear facility.

Two other major incidents at our laboratory during this early period, 1943–50, demonstrate the serious risk of accidents when working with ionizing radiation.[15]

One of these concerned a fire in the pile building. The aluminum-encased uranium slugs sometimes grew red hot and swelled, causing a blockage of the channels in the graphite through which cooling air was drawn. Then the blocked channels needed to be unloaded ahead of schedule. We accomplished this by shutting down the reactor, removing the shield plug of the blocked channel, and, with a long rod, pushing the uranium slugs out the back of the reactor, where they tumbled down an incline and rolled into a water canal maintained at a depth of 30 feet for shielding.

A thick wool mattress covered the incline to prevent damage to the aluminum-encased slugs as they fell from the back side of the reactor. One morning, after we had discharged a plugged channel, fire alarms sounded. The pile building reeked with the stench of burning wool. Soon smoke poured out of all the loading openings through the concrete shield on the front face of the reactor and through the experimental holes on the other sides of the reactor. Immediately, we realized that we had a tiger by the tail. If we cut off the fans, the lack of air would help smother the fire, but then the slugs in the reactor would become too hot. There was no way to apply CO_2 and other fire quenchers to the blanket without delay that could result in serious, and possibly fatal, human exposure.

The use of water presented the possibility of an explosive reaction with the burning slugs of uranium and the melting aluminum. Although the risk was great, there was no choice: We sprayed water onto the wool blanket and the back side of the reactor. We heaved sighs of relief as the fire stopped. No explosion, no longer a fire, and no one had been overexposed except for inhalation of some of the radioactive gases and particulates that escaped into the working area of the reactor building.[16]

The other serious incident in the very early period involved the release of small uranium particles that contained plutonium and fission products. These "hot" particles from a chemical separation operation would fall all over the laboratory area. Some of the cars in the parking lot contained scores of the hot particles, which also contaminated all the laboratory buildings.

Particles fell on workers' hair, and we located some on fallout trays as far as twenty miles away. Blotting paper on the bottom of these trays was removed each evening and taken to the health physics lab. Here in a photographic darkroom a large bare X-ray film was placed over each sheet of blotting paper and left there for twelve hours. The next morning the X-ray films were developed. Using a magnifying glass, we could see black starlike splotches all over the film. The central dark portion of a splotch was caused mostly by alpha particles from plutonium, while the "rays" were caused mostly by beta particles from fission products. This spectacle was a bit frightening to us because it revealed what was happening near such particles, many of which were in the lungs of our fellow employees and our own families living in Oak Ridge. When one would blow his nose, the handkerchief would become radioactive.

I persuaded Alvin Weinberg, the laboratory director as of 1948, to shut down chemical operations of the laboratory for a few days and to allow most employees to quit work early during an extensive laboratory cleanup period. Over my protests, he restarted the full laboratory operations a few days before I believed it was safe to do so.

SECURITY MEASURES CARRIED TO THE EXTREME

Security presented one of the annoying problems for professional employees in the early period at Clinton Laboratories. Almost everything we wrote required clearance before it could leave the laboratory. Sometimes measures reached what seemed an absurd level. Leading scientists had to use silly security names, such as "Mr. Rock" for Dr. Stone and "Mr. Holly" for Dr. Arthur Holly Compton.

One of the numerous instruments we developed could be moved across floors and desktops to measure alpha activity. To us it appeared to be snooping—like Pluto, the cartoon dog. So we named this instrument Pluto. As soon as security learned of this, they protested, saying that the name amounted to a giveaway that plutonium was being produced. We changed the name to Snoopy instead.

Even some of the physics lectures I had brought with me from Lenoir Rhyne College bore the "secret" stamp. Provoked, I sat down one day and in jest wrote "A Preliminary Report on the Low-Draft Fly Swat," a detailed report on how best to swat a fly. As required, I sent my report to laboratory records and the security office for evaluation. Sure enough, it came back as a top-secret document "affecting the national defense of the United States" (appendix 2). It was not declassified until January 17, 1957, more than eleven years later.

Most of us were not accustomed to having to obey so many rules and regulations. Complaints from the safety supervisor, R. B. Smith, peeved us. He inundated everyone, even Whitaker, who was director until 1946, with trivial demands. Smith objected because notices were fastened to the wall in Parker's office with cellophane tape. He was also upset when he observed two hickory nuts in Parker's ashtray.

Parker became so fed up he decided some horseplay was in order. He printed up small signs reading "The RBS Award, See Members of the Health Physics Division for Details." We placed the signs on bulletin boards throughout the laboratory late one evening. The next morning when employees arrived at work laughter echoed through the halls of the lab. Furious, Smith headed straight to Whitaker. Shortly afterward Parker received a phone call from Whitaker directing him to come to his office immediately.

When Parker arrived, he was told horseplay was not allowed in the laboratory and that public insults, such as the RBS Award, could not be tolerated.

Parker smiled and said, "Why, Dr. Whitaker, these signs have nothing to do with R. B. Smith. RBS stands for Rooms Bright and Shiny. Many of the rooms in the laboratory are very dirty, and it is well known in hospitals where I used to work with radium that such work must be done in rooms that are meticulously clean. They must be clean at all times and bright and shiny."

The two shared a hearty laugh. Unofficially Whitaker sympathized with Parker and the other scientists, but nonetheless the horseplay ended and the signs came down.

Whitaker left the laboratory after a year. A few years later he died of cancer. We never knew whether this malignancy originated the day he infringed upon health physics safety rules.

CHAPTER TWO

The Genie's Anger Unleashed

The Truman Administration's Greatest Mistake

I am become Death, the destroyer of the world.

VERSE FROM THE HINDU EPIC *BHAGAVAD GITA*,
RECALLED BY OPPENHEIMER AFTER HIROSHIMA

More than fifty years later, I remain firmly convinced that President Harry S. Truman and Major General Leslie R. Groves blundered in their decision to order atomic bombs to be dropped on Hiroshima and Nagasaki. Because timing is so critical, the following chronology should be helpful.

1939 Albert Einstein meets with Leo Szilard to discuss the development of an atomic weapon. Szilard (fig. 19), one of the discoverers of neutron emission of uranium, is highly regarded by Einstein.[1] As a result of their meeting, Einstein takes the liberty of approaching President Roosevelt via letter of August 2, with the prospect of developing an atomic bomb (appendix 1). Einstein's letter persuades FDR, and the secret project—later known as the Manhattan Project—is launched.

1942 The Chicago pile under the athletic bleachers of Stagg Field reaches criticality on December 2.

1943 The Oak Ridge graphite reactor reaches criticality on November 4.

1944 As early as spring of this year, a powerful Japanese group makes efforts to bring the war to a close. The group includes former prime ministers and others close to the emperor. Such influential leaders as Admiral Okada, Admiral Yonai, Prince Konoye, and Marquis Kido are in

Fig. 19. Leo Szilard, a student of Einstein who discovered the neutron emission of uranium and who believed the atomic bomb should not be dropped on Japan without a warning. (I signed one of the petitions initiated by Szilard.) (Photo: Ed Westcott, courtesy of DOE Photography)

the group. Successful in bringing about Tojo's resignation, they make Admiral Suzuki prime minister after Koiso's fall.[2]

On September 18 FDR and Churchill secretly initial an aide-mémoire at Hyde Park, New York (fig. 20, appendix 3). The two leaders believe efforts to develop the bomb "should continue to be regarded as of the utmost secrecy." They also agree that "when a 'bomb' is finally made, it *might* perhaps, after mature consideration be used against the Japanese, *who should be warned* that this bombardment will be repeated until they surrender."[3] The secret memo suggests the hesitancy of FDR in using the bomb at all. At a minimum he and Churchill contemplate warning the Japanese before repeating a bombing. When the existence of this document finally comes to the attention of Secretary of War Henry Stimson on June 25, 1945, it is explained that it has been "mislaid" by the Roosevelt administration.[4] There is no evidence that the joint memo by two of the twentieth century's greatest leaders is shown to Truman before he orders the bombing of Hiroshima and Nagasaki.

1945 Einstein writes Roosevelt again on March 25, introducing Szilard and essentially vouching for him (appendix 4). Einstein reports that Szilard "is greatly concerned about the lack of adequate contact between scientists who are doing this work and those members of your cabinet who are responsible for formulating policy."[5] The renowned scien-

Fig. 20. Two great leaders, Winston Churchill and FDR, who secretly initialed a memorandum at Hyde Park, New York, on September 18, 1944, in which they agreed that the Japanese should be warned before use of the atomic bomb and certainly before a second bomb. (Photo courtesy of Franklin D. Roosevelt Library)

tist urges FDR to pay heed to Szilard's "considerations and recommendations."[6] There is no evidence Truman was shown this letter before detonation of the first atomic bomb at Alamogordo, New Mexico, on July 16 or the bombing of Hiroshima on August 6.

FDR dies on April 12, never having had the opportunity to meet with Szilard. Truman takes office and within twenty-four hours is told about the Manhattan Project (fig. 21). Secretary of War Stimson advises him that the fission bomb is potentially "the most terrible weapon ever known in human history." Truman orders Stimson to form an interim committee to advise him on the use of the bomb in wartime.[7]

Hitler commits suicide April 30, as Allied forces surround Berlin. VE day occurs on May 8, when Germany unconditionally surrenders. Germany no longer presents the threat of development of an atomic bomb.

On May 25 Szilard calls on Matthew J. Connelly, secretary to President Truman, seeking an audience with Secretary of State James Byrnes. Three days later they meet and Szilard hands him a memorandum originally prepared for FDR.[8] The memorandum urges that the Japanese be given

Fig. 21. President Harry S. Truman, who took office in April 1945. When he became president, Truman did not even know of the existence of the Manhattan Project. (Photo: Ed Westcott, courtesy of DOE Photography)

the opportunity of a demonstration and hopefully a resulting surrender before the United States uses atomic weapons.

On May 31 the interim committee meets. Major-General Leslie R. Groves, controversial head of the Manhattan Project, questions the loyalty of "undesirable" scientists, such as Szilard, but says they can't be dismissed until after the bomb has been used or at least successfully tested (fig. 22). The meeting is chaired by Stimson, who "expressed the conclusion, on which there was general agreement, that we could not give the Japanese any warning." Stimson agrees that the most desirable target would be a vital war plant employing a large number of workers and closely surrounded by workers' houses.[9]

The interim committee meets again on June 1 to seek the views of "invited industrialists" from Westinghouse, DuPont, Union Carbide, and Tennessee Eastman. Tennessee Eastman operates the Y-12 plant that produces U-235, the explosive material that will be used in the Hiroshima weapon. Union Carbide operates the Oak Ridge K-25 gaseous diffusion plant that later produces tons of U-235. DuPont is the first operator of the four water-cooled graphite reactors at Hanford, which begins producing plutonium in late 1944 for the weapons later exploded at Alamogordo and Nagasaki. Secretary of State Byrnes recommends, and the committee agrees, that "the bomb should be used against Japan as soon as possible; that it be used on a war plant surrounded by workers' homes; and that it should be used without prior warning."[10]

Fig. 22. The controversial head of the Manhattan Project, General Leslie R. Groves. (Photo: Ed Westcott, courtesy of DOE Photography)

On June 16 a small science panel headed by nuclear physicist J. Robert Oppenheimer acknowledges that scientists working on the project have different views regarding a demonstration but concludes that there is "no acceptable alternative to direct military use."[11] However, the panel is unable to reach "sufficient agreement . . . to unite upon a statement as to how or under what conditions such use [is] to be made."[12]

Twenty days before we secretly test the plutonium weapon, the Japanese emperor, during the Imperial Conference on June 26, states that he desires the development of a plan to end the war.[13] This is followed by a renewal of efforts by Japan to persuade the Soviet Union to intercede with the United States. On July 13 the results of a poll of 150 scientists on the Manhattan Project are made known to Arthur Holly Compton. Compton, who suggested the poll, is informed that 83 percent (including me) favor some type of "demonstration" of the enormous power of the bomb, which would serve as an inducement to Japanese surrender and avoid countless unnecessary deaths.[14]

The Trinity Test of the first atomic bomb occurs three days later at Alamogordo, New Mexico, on July 16—more than two months after Germany's surrender (figs. 23, 24; see also preface). The light from the blast is visible nearly two hundred miles away. We now know that what has happened on the deserts of Alamogordo might some day destroy our big cities and obliterate the lives of many thousands of human beings. The number of lives lost from blast and heat will be equaled or exceeded by the number lost over time from radioactive fallout.

Realizing the horror this weapon of destruction could bring to human life, a plea goes from many of the scientists at the various laboratories to President Truman to use this weapon with caution and display its potential on some island off Japan rather than on a Japanese city. Within a few days after Trinity, Szilard submits a petition signed by sixty-seven scien-

Fig. 23. Preparation for the Trinity Test: The world's first atomic bomb is loaded on the Trinity Tower in the New Mexico desert, to be raised 100 feet in preparation for its detonation on July 16, 1945. (Photo: Ed Westcott, courtesy of DOE Photography)

Fig. 24. J. Robert Oppenheimer (left) and Major General Leslie R. Groves, viewing what remained of the base of the Trinity Tower after the blast. (Photo: Ed Westcott, courtesy of DOE Photography)

tists calling for a demonstration.[15] Compton, a member of the science advisory panel to the interim committee, forwards the petition to the military along with the opinion poll. Compton states: "You will note that the strongly favored procedure is to 'give a military demonstration in Japan, to be followed by a renewed opportunity for surrender before full use of the weapons is employed.' This coincides with my own preference, and is, as nearly as I can judge, the procedure that was found most favored in all informed groups where the subject has been discussed."[16]

On July 25, nine days after the frightening test at Alamogordo, Colonel K. D. Nichols (fig. 25), who serves under Major-General Groves, authors a memo to Groves recommending that the views of the scientists—as expressed in the opinion poll and the various petitions—"be forwarded to the President of the United States."[17] Leslie Groves and others decide not to pass on the views of those of us who worked on the project. It is undisputed that Groves's dislike of Szilard is so intense that he has him constantly under surveillance. Groves even goes so far as to draft a memo

Fig. 25. Colonel K. D. Nichols, who served under Groves and who wrote a memo nine days after the Trinity Test recommending that the views of the scientists as expressed in the various petitions "be forwarded to the President of the United States." (Photo: Ed Westcott, courtesy of DOE Photography)

in the name of Secretary of War Stimson directing that Szilard be interned for the duration of the war.[18] Almost anything Szilard favors, Groves opposes, and vice-versa. Szilard, the most enterprising and persistent of those urging that we not let Hitler develop the first atomic bomb, thus becomes the wrong champion for a noble cause.

Secretary of War Stimson sends "urgent" message no. 41011 to Truman on July 30, asking permission to drop the bomb. Truman responds: "Reply to your 41011 suggestions approved. Release when ready but not sooner than August 2. HST" (appendix 5).[19] President Truman is attending the Potsdam Conference and does not depart Germany until August 2 (fig. 26). The materials sent to Groves (which express the views of the vast majority of us, that a demonstration should precede military use) are not forwarded to Truman until after the order to drop the bombs has been given.

On August 6, at 8:15 A.M., a B-29 named *Enola Gay*, piloted by Major Thomas W. Ferebee of Maxwell, North Carolina, drops "Little Boy" over Hiroshima (fig. 27). Within a few short days more than a hundred thousand Japanese die from the first atomic weapon used to destroy mankind. Some of them leave only their shadow burned on the sidewalks. Almost 98 percent of Hiroshima's buildings are destroyed or severely damaged by the uranium-235 gun assembly bomb that all of us helped develop. Within

Fig. 26. Truman with Churchill at the Potsdam Conference, where he received Secretary of War Stimson's urgent message No. 41011 asking permission to drop the atomic bomb on Japan. Truman authorized use of the bomb, "but no sooner than August 2, 1945," when he was scheduled to depart Germany. (Photo courtesy of Harry S. Truman Library)

Fig. 27. "Little Boy" (top), a gun-design atomic bomb using enriched uranium produced at Oak Ridge, and "Fat Man," an implosion-design atomic bomb using plutonium produced at Hanford. Little Boy was dropped on the city of Hiroshima on August 6, 1945; Fat Man was dropped on Nagasaki on August 9. (Photos: ORNL)

five years two hundred thousand Japanese are dead from the immediate and delayed effects of this single weapon.

The third nuclear weapon developed so far bears the nickname "Fat Man" and is a duplicate of the Trinity Test bomb that used plutonium-239 instead of U-235. Incredibly, the decision of when to drop this bomb is "left entirely to Groves."[20] On August 9 a B-29 named *Bock's Car* drops

"Fat Man" over Nagasaki (fig. 27). Half the city is instantly destroyed, and of the approximately 270,000 inhabitants, 70,000 are dead by the end of the year. If one adds latent deaths from these two bombs, the total number of Japanese killed reaches half a million.

Before Germany's surrender, I almost fanatically desired that the bomb would meet all our expectations. Today I wish it had been a failure. Individuals within the Truman administration stonewalled our petition. It never reached the president before he made the decision to use the bomb. There is no evidence that Truman read Roosevelt's secret agreement with Churchill that stated the Japanese should first be warned.

Truman's personal journal contains evidence that he favored a warning. On July 25, 1945, he wrote that the "target will be a purely military one and we will issue a warning statement asking the Japs to surrender and save lives."[21] He did issue an ultimatum for unconditional surrender, and pamphlets were dropped over Japanese cities. But the pamphlets gave no indication of the unprecedented horror of the threat. The Japanese quite reasonably interpreted the threat as meaning that they could expect a continuation of conventional bombing.

The bomb we dropped over Hiroshima was a U-235 weapon. Both Byrnes and Stimson strongly favored bombing without warning. Motivated by the expectation that a nuclear monopoly would prove to be a "diplomatic master card" in U.S. relations with the Soviet Union, the United States took the lives of hundreds of thousands of Japanese—mostly women and children—to add emphasis and strength to our master card.[22] However, the monopoly lasted for only about a year. The stigma of what we did can never be erased.

Anxious to determine the comparative destructive capacity of a plutonium weapon in terms of human lives, the U.S. military saw Nagasaki as an opportunity to determine how the plutonium weapon rivaled the uranium device. As a consequence, on August 9, 1945, the bomb dropped over Nagasaki only repeated the tragedy. The plutonium bomb packed even more destructive power than the U-235 weapon, but the number of fatalities in Nagasaki was somewhat less than at Hiroshima because the hills of the city provided effective shielding in some areas.

This unwarranted act reflected the callous disregard for the lives of civilian men, women, and children on the part of U.S. military advisers. Determined to obtain a "kill count" of plutonium as a weapon for comparison to the uranium weapon, those in control did not want a surrender until the two weapons could be comparatively "field tested." By August 10, Truman put a stop to the military's plan to drop yet another atomic bomb on Japanese soil, in this case Tokyo. He stated to his Cabinet, "The thought of wiping out another 100,000 [is] too horrible."[23]

Japan surrendered on August 14, 1945, five days after Nagasaki. Often the excuse is given that the dropping of the two bombs over Japanese cities may actually have saved human lives, especially the lives of U.S. soldiers. One or two demonstrations of the destructive power of the atomic weapon over an uninhabited Japanese island for Japanese observers would have ended the war just as soon and saved hundreds of thousands of lives, most of them civilians.

No justifiable reason existed to bomb Nagasaki. Japan wanted peace. On July 7 the Japanese emperor had asked Russia to act as an intermediary to bring about peace. Since the Japanese military had fostered the concept of complete obedience to the emperor, it could not effectively rebel.[24] The emperor desired peace.

Immediately after the atomic bombing of the two Japanese cities, scores of scientists who had actively participated in the development of the atomic bomb publicly expressed their outrage. We emphasized that our country could not possibly possess a monopoly in the development and use of this terrible weapon. We argued that the only hope for the future was to work for a lasting world peace.

A group of us from Oak Ridge traveled to various cities giving lectures and calling for the strengthening of the United Nations to enable it to maintain world peace. Senator Estes Kefauver (1903–63) served as our political leader and proved to be a forceful one (fig. 28). Kefauver, from Tennessee, offered bills to modernize Congress and wrote an article titled "Twentieth Century Congress" that attracted much attention. He strongly supported efforts to create the United Nations. A world statesman before his time, he attempted to strengthen the United Nations and served as a major catalyst for the World Federalist Movement.

I lectured in several cities and met with some persons of world renown to discuss strengthening and restructuring the United Nations. I emphasized that a nation's partial loss of sovereignty would represent a negligible price to prevent the entire loss of civilization as we know it.

The two most famous scientists with whom I met in these discussions were Niels Bohr (fig. 29) and Albert Einstein, recipients of the Nobel Prize in physics in 1922 and 1921, respectively. I talked at length with Bohr one evening in Copenhagen around 1947. He favored strengthening the United Nations but expressed skepticism regarding the possibility of accomplishing this in our lifetime.

Fig. 28. Senator Estes Kefauver (left) of Tennessee, who in the postwar period served as our political leader to strengthen the United Nations. A major catalyst for the World Federalist Movement, Kefauver believed lasting world peace was attainable. (Photo: Ed Westcott, courtesy of DOE Photography)

Fig. 29. Danish nuclear physicist Niels Bohr, a Noble Prize winner, who spoke with me at length in Copenhagen after the war about how to strengthen the United Nations. (Photo: Ed Westcott, courtesy of DOE Photography)

Einstein invited me to his home in Princeton in the winter of 1947. In this simple frame structure at 112 Mercer St., we sat before his open fireplace and discussed the urgency of getting our message to the world community before it was too late. Einstein, like Bohr, had a warm, engaging personality.

Both Bohr and Einstein made me forget my insignificance in the presence of two of history's greatest minds. Intensely interested in learning, they questioned me about chronic damage from ionizing radiation and the strides we had made in health physics.

My knowledge on such subjects as radiation-induced cancers, cataracts, genetic mutation, life-shortening, and teratogenic damage was pitifully limited at that time, but I responded to their questions as best I could. They also expressed interest in health physics as practiced at Oak Ridge and how we managed to keep radiation exposures so low in these dangerous operations. Both men were a bit younger than my father, gray-haired and gentle in their manner, such that they reminded me of him.

The World Federalist Association espoused our goals of a peaceful, liveable, global environment maintained through the development of enforceable world law, but unlike us it did not advocate a strong U.N. military force. Most of its members wanted this but realized it would have to come much later. The association called for a U.N. Environmental Security Council with power to develop and enforce international environmental regulations. A strengthened international Court of Justice would adjudicate environmental disputes.

Perhaps in the early postwar period we behaved like novices, setting our sights too far ahead. Helen and I learned in our many hikes in Switzerland that when the path got narrow around a ledge with a steep dropoff of a thousand feet, we had to look ahead, not down, and take short steps—or we invited disaster.

CHAPTER THREE

The Early Years at Oak Ridge National Laboratory

Every problem has in it the seeds of its own solution.
NORMAN VINCENT PEALE

FINDING SOLUTIONS TO COMPLEX PROBLEMS

In September 1943 I, along with four other forerunners of the new profession of health physics, arrived at Clinton Laboratories shortly before the pile first became critical at 5 A.M. on November 4, 1943. We faced a mind-boggling challenge: responsibility for the largest single concentration of radioactive sources in the world. Since none of our human senses could detect even fatal doses of radiation, a functional health physics program at what became Oak Ridge National Laboratory (ORNL) was imperative. Atomic energy presented health threats to us, our families, and future generations that were never imagined before.

We detected airborne radioactivity with a Constant Air Monitor that sucked air through filter papers or by use of electrostatic precipitators that collected radioactive dust particles on sheets of aluminum foil. Our group monitored personnel by pencil meters clipped on a worker's shirt or a laboratory coat. By February 1944 we used film badges. In sum, we developed a system of radiation protection that combined personnel monitoring, laboratory survey, area monitoring, and portable as well as fixed survey equipment of various types. Collection and analysis of urine samples of radiation workers occurred routinely. "Protective clothing" was always available, if needed. It provided no useful shielding but did prevent radioactive materials from collecting on the body, and in the case of visitors it helped prevent the possiblity of contaminated clothing leaving the laboratory.

By the 1950s little "doghouses" sprang up in many locations in the laboratory and in twenty outside locations to a distance of fifty miles. These "doghouses" contained an assortment of continuously operating radiation detection equipment. We also collected water samples from lakes, rivers, rainwater, and nearby springs and wells up to a distance of about twenty miles.

In 1955 we built our first "Whole Body Counter" (WBC), a large device that looked a bit like today's medical scanning instruments for magnetic resonance imaging and computerized axial tomography (MRI and CAT scanners). Our WBC measured levels of ingested and inhaled radionuclides in the gastrointestinal tract and the lungs. It also determined levels of radionuclides accumulated in body organs.[1]

When we built the WBC, we experienced an unanticipated obstacle. In order to reduce the background level from cosmic and terrestrial radiation, we planned for all sides to be constructed of 6-inch-thick iron plates. However, every test we conducted of sample iron plates revealed that the plates themselves were radioactive.[2] Fortunately, we located a vendor who was able to retrieve iron plates from a ship sunk before the atomic age. To our delight, the natural background in these proved to be very low.

We mistakenly believed we had resolved the issue. Our original plan was to have the worker/patient remain stationary and lie flat while scanning euipment fastened to a trolley on the ceiling moved along a track directly above the individual being scanned. However, the workmen had unknowingly magnetized iron plates that moved along the track by using direct-current welding equipment. Whenever the magnetized iron moved, it changed the background reading, resulting in unreliable scanning results. We knew of no way to demagnify the iron plates except to heat them at high temperatures, which was not practical.

Now, a new problem appeared, since we could not follow our initial plan to move the scanning equipment on an overhead track. Fortunately, the room was large enough to permit us to alter our plans and mechanically move the person (rather than the equipment) lengthwise past the now stationary overhead equipment.

Analytical chemistry problems proved challenging. For example, methods had to be developed to analyze urine, feces, fish, vegetation, and dust for numerous types of radioactive contamination. Ralph Firminhac and Larry Farabee took the lead in these early developments.

We health physics surveyors gained the highest respect and confidence of the regular laboratory employees, outside contractors, and especially the union workers. The riggers, pipefitters, mechanics, and reactor engineers trusted us and depended upon us. It would be easier to convince a skydiver to leap from the plane without a parachute than to get them to begin an operation without the health physicist at arm's length, instruments in hand. Workers literally refused to begin or stay on a job unless the required number of health physics surveyors provided ongoing surveys to limit the time of exposure. (In some instances they were even instructed to run if high levels of radiation suddenly developed.)

Caution ruled every step. Workers in a pipe tunnel might be leisurely working on some minor repairs or turning certain valves where the radiation level was only about 8 mr per hour, a thousand times the initial Oak Ridge natural background, when suddenly the radiation level would jump to 8 R per hour, a million times natural background, due to "hot" solution passing down a nearby pipe to the underground waste storage tanks. In a sense, the health physicist with his instruments served as the eyes and ears for those who worked with this dangerous and invisible energy source.

During our first few years of operation, several spills of radioactive materials occurred. Some of the scientists and health physicists, all of whom were white, considered cleanup a task for their "servants"—the African Americans.

Careless former college professors presented the major problem. Some showed disdain for health physics restrictions. Ironically, the trust and high respect we received from blue-collar workers sometimes failed to transfer to certain scientists. Either ego or pride caused them to take offense when we tried to protect them.

I persuaded the managers of areas I was not directly responsible for (such as nuclear engineering, physics, biology, and chemistry) that each scientist or engineer responsible for a hot spill, whether a director or Nobel laureate, needed to clean up the mess himself while our health physicist, with Geiger counter clicking in hand, supervised at a safe distance. After this new regulation, spills of radioactive material rarely occurred in the laboratory.[3]

We did experience a serious chemical explosion on November 20, 1959. About midnight, my senior applied health physicist, Roy Clark, called to awaken me. My adrenalin kicked in when Roy reported that there had been a big explosion in Building 3019, spreading contamination into the street and adjacent buildings. Fortunately, he said, it seemed that no one had been hurt.

I dropped the phone, slipped on my clothes, and drove to the laboratory from my home as fast as my car would take me. After pulling on protective clothing, I grabbed a face mask, a Hurst neutron dosimeter, a Geiger counter, and an alpha survey meter, and rushed to the accident scene.

Before I had even reached the contamination site, my alpha counter exceeded the low-scale maximum limits,[4] but my Geiger counter was reading only 15 mrad per hour. These readings strongly suggested that the explosion had scattered plutonium in a wide area of the laboratory, but the low reading of beta and gamma radiation of my Geiger counter indicated that a criticality accident was not taking place. To make certain, I turned on my neutron dosimeter and was relieved when it read zero. Clark roped off a wide area around the south side of the building where the explosion had occurred. As I approached the security rope, my alpha counter exceeded the high scale of 1 million disintegrations per minute. Moments later it read zero. Now I knew that I was in a very high alpha radiation background area because the alpha counter was reading zero in an extremely high background area. A proportional alpha counter and a Geiger beta-gamma counter responded only to pulses of radiation (individual photons or ionizing particles), so if the pulses arrived very close together, the counter failed to recover during pulses; consequently there were no counts—that is, only a DC current in the counter—and the index needle of the counter dropped to zero.

Thank God, my Hurst neutron dosimeter, developed by my colleague G. S. (Sam) Hurst, read zero (fig. 30).[5] Since there were no neutrons, I knew that, at least at that moment, no assembly of plutonium was occurring sufficient to reach criticality and instantaneously deliver a fatal dose to anyone nearby. The presence of neutrons would indicate a criticality accident still in progress that could deliver a fatal dose of radiation in minutes of exposure.

Clark, the most outstanding health physics surveyor with whom I had ever worked, had everything under control. People close in were wearing face masks and protective clothing.[6] The fire truck stood by on alert. Clark had roped off the area above 10,000 alpha disintegrations per minute, or the top of the low scale on the alpha survey meter.

Trucks already on their way soon spread tar over the contaminated road and ground just outside the broken door leading into the building where the explosion had occurred. The tar held down the plutonium so that it would not blow over the laboratory. Within hours, workmen had sprayed

Fig. 30. Sam Hurst (left) and Rufus Ritchie, two of the most talented and reliable scientists who worked under me at ORNL. (Photo: ORNL)

paint on the outside of all the nearby buildings that had been contaminated with plutonium.[7]

Eight hours later, after helping to stabilize the situation, I took several hot showers, scrubbed down, and shampooed my hair again and again so it registered below the acceptable alpha counter limit of 500 disintegrations per minute. Although exhausted, I felt relieved that we had avoided any fatalities. It was now 8 A.M., and I was hungry, so I walked to the laboratory cafeteria.

On the way, Dr. Alvin Weinberg, the laboratory director, arrived for work and overtook me (fig. 31). He asked, "How did things go last night, Karl?"

I replied, "We had a hard time. Clark did a magnificent job. We are all exhausted from the ordeal but relieved that things are under control."

Then Weinberg asked, "What instruments did you take to the scene of the accident?"

I replied, "A Geiger counter to measure the beta-gamma activity, a proportional counter to measure alpha, and a Hurst dosimeter for the neutrons."

Weinberg gave me a friendly pat on the back as he laughed and said, "Karl, surely you did not expect to measure any neutrons?"[8]

Fig. 31. Al Weinberg (right), my laboratory director, at a 1968 ceremony honoring my twenty-five years of outstanding scientific service at ORNL. (Photo: ORNL)

I answered, "Al, I knew there was a lot of plutonium in this building. I don't take any chances with criticality."

Some months later workers dismantled the tar road with jackhammers and hauled it off to our hot burial ground. Disassembly of the building just across from the scene of the accident followed. The workers tore down the structure section by section, placed its pieces in plastic bags, and hauled them off for burial. Samples of the residue drained from the ruptured tank that had caused the explosion indicated that enough plutonium existed to produce at least one critical assembly. I still soberly remember the laugh from Weinberg, who had never even considered criticality a possibility. Sometimes my stubborn conservatism has paid more dividends than any intelligence I could hope to possess.

Complacency can never be permitted when one deals with atomic energy; an occurrence at the Y-12 plant demonstrated this. I was not responsible for health physics in this plant and seldom visited it. However, I knew the layout of the buildings and the operations taking place there. In one of the buildings, scientists worked with solutions of highly enriched uranium-235—concentrations of U-235 greater than 95 percent (natural uranium has a concentration of only 0.719 percent).

The Y-12 health physics group had probably cautioned everyone in this operation—except the janitor—not to bring into the building any containers that might be deep and wide enough to permit a critical assembly.

Any vessel brought into the area had to be small and had to have a large outside area relative to volume in order to permit the escape of neutrons and avoid a critical assembly of fissile material—in this case, U-235. Not infrequently, years of routine operations cause people to become a bit careless and to forget that their hands grasp the tail of a sleeping tiger.

Even a half-gallon jar could contain a quantity of U-235 that bordered on high-risk. A 5-gallon container in the building should have been forbidden.

One morning the janitor commenced his early morning tasks in the Y-12 building before the operators arrived. Annoyed that a puddle of dirty yellowish solution had repeatedly collected on the floor, he "solved" the problem. He retrieved a 55-gallon rain barrel from outside the building and placed it under the pipe where it would catch the slowly dripping fluid. Day after day and week after week this barrel remained in an inconspicuous place behind some machinery.

No one in the Y-12 operation apparently considered the black janitor significant enough to be part of an informed operation. On the morning of June 16, 1958, a sharp bang from the 55-gallon drum and a blinding blue flash of light caused everyone in the building to immediately rush for the nearest exit—a criticality accident was under way.

I was in my office at X-10 that morning when the phone rang. I picked up the receiver to hear someone shouting, "We have a criticality accident at Y-12 and thousands of employees are evacuating the plant!"

I reached for my emergency kit and rushed for the door. My assistant, Hubert Yockey, grabbed his kit as well, and we ran out to a company car. I drove the ten-mile distance over a rough sandy road to Y-12 in eight minutes. Hundreds of persons milled outside the gate. Only our car was permitted past the guards and allowed to enter the area where minutes before thousands of people had been at work.

When Yockey and I entered the windowless building that contained the problem, darkness engulfed us. I muttered to myself, "My kingdom for a flashlight." (After this, flashlights became an essential part of all our emergency kits.) A faint light shone from a battery-operated emergency lamp in the far end of the building, and we "homed in" on the life-threatening barrel as best we could. Unable to read the scales on the Geiger counter, we could hear the clicks sounding faster and faster as we approached the far end of the building.

Each time the Geiger counter needle banged the end of the scale, we changed to a higher scale as we approached the radioactive source. Most important, we listened for clicks on our Hurst neutron dosimeter. Fortunately, we heard none. The presence of neutrons would mean a life-threatening critical assembly still existed. At any moment the solution might settle down and reach a critical configuration producing another ominous brilliant flash of light and a deadly burst of neutrons and gamma rays.

The clicks from our Geiger counter saturated or ran together on the highest scale, so the counter stopped clicking.[9] The dose rate was so high—above 500 R per hour—that it had blocked the counter. We ran from the building.

Yockey and I drove quickly to the security gate and told the waiting engineers that there was no longer a critical assembly in the barrel, but that they must be very careful and spend only seconds as they attempted to "defuse" the barrel because it was very radioactive.[10] They put on protective clothing and masks, rushed into the building, and poured into the barrel a high concentration of borax, which absorbs neutrons and "kills" any possibility of a critical assembly of the fluid. The barrel could be removed in a few hours because of the short half-life decay of the radionuclides.[11]

Minutes later, with the help of Y-12 health physicists, we rounded up all the employees who had been in the building at the time of the accident. We required them to shower immediately and scrub repeatedly. The Y-12 medical doctors provided me with 5 cubic centimeters of blood from each of eight highly exposed workers. They added a few drops of heparin to each blood sample to prevent coagulation. I took the blood samples to our low-background counting facility, where we analyzed them. We determined that each had received an impermissibly high neutron and associated gamma dose (see table).[12]

The lax health physics regulations at Y-12 contrasted sharply with our X-10 facility. None of the eight persons was wearing a personal dosimeter at the time of the accident.

The doses were determined by Hurst and his group, and given in energy absorption units of rads. The proper value in damage units or rems depends on the type of damage considered. Today, the International Commission on Radiological Protection (ICRP) recommends a value of Q of 30 for fast neutrons. Five of the Y-12 workers experienced radiation sickness and epilation (loss of hair). Those who received 365 and 236 rads of radiation experienced some hemorrhaging. The five with the highest doses suffered considerable epilation. Even the individual who received only 22.8 rad showed some symptoms of radiation injury.[13] Of the eight Y-12 employees

Dose Estimates of Y-12 Employees during the Criticality Accident on June 16, 1958

	Estimates of Total Body Dose				
Employee	Thermal Neutron Dose (rad)	Gamma Dose (rad)	Total Dose (rad)	Symptoms	Total[a] (rem)
1	96	269	365 N	D, H, E	749
2	89	250	339 N	D, E	695
3	86	241	327 N	D, E	671
4	71	199	270 N	D, E	554
5	62	174	236 N	D, H, E	484
6	18	50.5	68.5 N	A	141
7	18	50.5	68.5 N	A	141
8	6	16.8	22.8 N	S	47

Key: N = Not wearing radiation dosimeter at time of accident.
D = Symptoms of radiation damage
H = Hemorrhage
E = Epilation

Note: Dose estimates made by G. S. Hurst. From Morgan and Turner, *Principles of Radiation Protection*, table 1-11, p. 44

[a] Values were not given originally by Hurst in rem units but are added here by K. Z. Morgan using a quality factor Q = 5 for thermal neutron dose. The thermal neutron reaction takes place with the sodium in the blood and is ${}^{23}_{11}\text{Na} + {}^{1}_{0}\text{n} \rightarrow {}^{24}_{11}\text{Na}$ (14.96 h).

listed in the table, three died of cancer and three others were diagnosed with cancer. One died of a stroke, and one had no major health problems at least as of 1995, when I last checked. Perhaps he was the one with the lowest dose, or he may have possessed the most efficient immune system.

Yockey and I incurred a radiation exposure dose of about 5 rem during this emergency. The Y-12 accident had a sobering effect on me. I tightened even further the radiation protection measures at X-10. I thank God that we never experienced a criticality accident in the X-10 area where I was responsible.[14]

INTERESTING VISITORS

One of the pleasantest experiences of my employment at ORNL was the constant stream of visitors from all parts of the world. Some of the more

Fig. 32. Then Senator John F. Kennedy and Jackie Kennedy visiting ORNL shortly before he announced his candidacy for the presidency. (Photo: ORNL)

famous visitors included Senator John F. Kennedy and Mrs. Kennedy (fig. 32), President Eisenhower, Eleanor Roosevelt, Admiral Rickover, and King Hussein.

In 1947 L. H. (Hal) Gray telephoned from London to find out if I could arrange an unscheduled visit for him to my laboratory (fig. 33). Gray was one of the best-known medical physicists of the prewar period, best known for his development, along with W. H. Bragg, of the Bragg-Gray principle.[15] With strict security, it often took weeks to obtain a clearance, especially for a foreign visitor. To obtain his clearance, I called my close friend Shields Warren, director of the Biology and Medicine Division of the Atomic Energy Commission in Washington. Four days later, after the propeller plane hop-and-jump flight—London to Shannon, Ireland; Shannon to Gander, Canada; Gander to New York; and New York to Knoxville—Gray visited my laboratory. Our research projects excited him, and he offered some useful suggestions.

That evening Gray and his wife joined Helen and me in our home for dinner. Mrs. Gray had been blind during their entire married life, but she refused to allow her disability to present more than a minor handicap. The fragrance of the roses in our garden and the songs of the mockingbird, cardinal, song sparrow, Carolina wrens, and catbirds thrilled her.

Fig. 33. L. H. "Hal" Gray (second from right) on a visit to ORNL in 1960, with Hubert Yockey (left), Bob Birkhoff, and Jim Hart (right) of our group. (Photo: ORNL)

A special treat was Rolf Sievert's visit from Sweden in 1957 (fig. 34). I visited Sievert many times in Sweden because of my work as twenty-year member of the main commission of the ICRP. A top medical physicist, he developed the first Whole Body Counter. He chaired the ICRP from 1955 to 1962. Sievert, who considered me the leading authority on health physics in the United States, visited Oak Ridge to confer about instruments we had developed and to anticipate problems surrounding the new reactor that was being constructed in Stockholm. A big man in more than one sense, Sievert must have weighed 300 pounds. But every pound he gained during the years I knew him seemed to enhance his sense of humor. With no first-class sections in the aircraft of those days, and the seats just as narrow as tourist class today, Sievert always required two seats on the plane.

Val Mayneord and Greg Marley of the United Kingdom also visited ORNL around 1958. Marley was the leading health physicist in Europe, while Mayneord was recognized as an authority on dosimetry of radiation.

Sir John Cockcroft visited various nuclear facilities including our laboratory during the 1950s (fig. 35). On these visits he learned firsthand of the problems we encountered with radioactive particulates, as well as noble

Fig. 34. Swedish scientist Rolf Sievert, one of my favorite visitors to ORNL. (Photo courtesy of World Health Organization)

gases argon, xenon, and krypton, that discharged in large quantities up the stack from the Oak Ridge AGR. Noting that exhaust air from our reactor passed through filters to remove particulates before its discharge up the stack, he decided to make major modifications in a plutonium-producing, air-cooled plant under construction at Windscale, England. This plant, like the Oak Ridge reactor, was an AGR. But upon returning home, Cockcroft found that the Windscale stack already reached 50 feet up. Undeterred, he ordered what later became known as Cockcroft's Folly, the addition of a filter house on the stack of the plutonium-producing graphite reactor.

STIMULATING TALENT

Otto von Stuhlman, who had been my professor at the University of North Carolina, and Walter Nielsen, my adviser at Duke, imbued me with a belief in the importance of research and of understanding how and why things occur. With this in mind, I brought into my health physics program talented young scientists bursting with new ideas. Before long our group was publishing more important papers in the scientific literature, producing more new instruments, and obtaining more patents than other, less innovative groups.

In 1947 members of the physics and engineering divisions jealously complained to the acting laboratory director, Eugene Wigner (fig. 36),

Fig. 35. British scientist Sir John Cockcroft, who visited our laboratory during the 1950s and subsequently caused major modifications to be made at the Windscale plant. (Photo courtesy of UKAEA)

Fig. 36. Eugene Wigner, winner of the 1963 Nobel Prize in physics and acting laboratory director at ORNL until 1948. (Photo: ORNL)

claiming that health physics should be confined to applied work. They asked Wigner to break up and redistribute my division.

The issue heated up so much that it became the principal subject of discussion at one of the weekly meetings of laboratory directors. Arguments lasted for almost two hours. Two-thirds of the directors at the meeting opposed me and favored the breakup of the Health Physics Division.

When the meeting ended, I walked out with Wigner. As we left the room together, he put his arm on my shoulder in a fatherly manner and said, "Karl, it's a shame you could not agree with us." I felt discouraged. To my surprise, I learned a few days later that Wigner completely agreed with me.

Both Wigner and Alvin Weinberg, who took Wigner's position in 1948, not only tolerated but sought employees who had the guts to disagree with them. They did not behave like so many other directors who only want to look in the mirror and see a reflection of their own views. It was indeed fortunate for me and for health physics that I worked under such directors.

CHAPTER FOUR

My Biggest Mistake

This above all: to thine own self be true,
And it must follow, as the night the day,
Thou canst not then be false to any man.

WILLIAM SHAKESPEARE, *HAMLET*

During the twenty-five-year period following World War II, the nuclear industry mushroomed into a multibillion-dollar enterprise with virtually unlimited resources and immense political influence. By the late 1960s and early 1970s I was feeling increasing concern about the newly developed liquid metal fast breeder reactor (LMFBR). I remained firm in my conviction that the molten salt thermal breeder (MSTB) that we had developed and built at ORNL provided a safer and more acceptable means of nuclear power. The MSTB held great promise as an energy system for the future and received enthusiastic support among most scientists at ORNL.

I arranged to deliver a paper on the dangers of the LMFBR at an international scientific meeting of radiation physicists in Neuherberg, Germany (near Munich), on July 5, 1971. I intended to express my view that the LMFBR offered a relatively easy means of access to an atom bomb and that I much preferred the MSTB. It was frightening to think of tons of plutonium as spent fuel from reactors being shipped through New York and other big cities to processing plants, then to fuel fabrication facilities, and finally back to LMFBRs all over the world.

In the Neuherberg paper I pointed out that plutonium-239 served as the operating fuel in the LMFBR and would be bred in relatively large concentrations in the natural uranium, U-238. By means of a relatively simple procedure one could separate the plutonium and construct a low-level atom bomb. This "poor boy" atomic bomb could still possess roughly the capacity of the Nagasaki bomb. With thousands of tons of LMFBR fuel being shipped from these reactors as they were being built in different parts of the world, enough of this fuel might be accumulated by terrorists to construct several such weapons. Only one ton of LMFBR fuel would furnish enough plutonium in the hands of an enemy country to wipe out a city the size of Jerusalem or Tel Aviv. Devastating blows could be delivered to larger cities, such as New York or Los Angeles.

Published scientific papers demonstrated that if a developing country paid a high price to employ trained nuclear scientists and engineers, sufficient Pu-239 could be separated to build an atomic bomb within a month. Before the LMFBR fuel was received, it would be necessary to provide a well-equipped, underground, silolike laboratory for the trigger and machined parts for the weapon. The size and weight of the bomb would depend mostly on the skill of the engineers in developing the implosion device and the neutron-producing trigger for the weapon.

As I prepared my paper, I asked myself what the president of the United States would do if he received a note announcing, "An atomic bomb is hidden in Washington and another in New York City. Both have electronic fuses. Unless you meet our demands by midnight, we will eliminate these cities."

The plutonium-239 produced by the LMFBR not only would serve as an enticement to terrorists, it also presented one of the greatest hazards of all radioactive materials.[1] I shuddered to think of tons of plutonium being shipped all over the world. U-233 held much less appeal for terrorists since it can be denatured (rendered impotent) by using depleted uranium (as is the case with U-235) and can be rendered unsuitable for use in bombs. This was not the case for Pu-239.

I intended to point out that our MSTB with its U-233 presented a more difficult target for terrorists.[2] In attempting to construct an atomic bomb with U-233 from an MSTB operation, terrorists would experience enormous difficulty from the intense gamma ray background that made remote-control operations necessary.[3] They would also have to contend with spontaneous fission with neutron emission that would cause predetonation unless the weapon was detonated instantaneously by implosion.[4] U-233 and its daughter products produce intense X-rays. Like Pu-239 and U-235, U-233 also decays spontaneously part of the time by fission, emitting neutrons that could cause predetonations.[5] The terrorists would be forced to take extreme precautions, using sophisticated remote-control equipment, in working with U-233.

Another proliferation restraint and an extremely important safety feature of the MSTB was a unified system for continuously removing the fission products, incorporating them in glass rods, encasing these in stainless-steel jackets, and shipping the assemblies for permanent disposal in an abandoned salt mine, such as the Carey and Hutchinson salt mines in central Kansas. We had researched these for several years. Agents of the International Atomic Energy Agency would be present for all change-of-fuel operations to assure that they were conducted as specified and to prevent diversion of fissile material. During such operations uranium rods are pulled with remote control devices and placed in lead containers. Much of the material I planned to present at Neuherberg was not detailed in my written speech, but was included in my notes for the long discussion period following my presentation, where we would go into details.

From an economic point of view as well, the LMFBR had problems. For example, it was extremely expensive to separate isotopes, such as pure lithium-7. Further, the extraction of protactinium-233 (Pa-233) from molten fluorides by liquid bismuth was not only costly but also complicated. For all these reasons, the LMFBR was not acceptable to me.

When I completed and submitted my paper, I sent it through the internal review process and agreed to the few suggested changes. I made certain not to reveal classified information and, as requested, shipped 250 copies to the meeting chairman in Neuherberg for advance distribution to those who would be attending.

Before the conference Helen and I went on vacation. We decided to revisit Zermatt by way of Zurich. The train ride to Zermatt took us through gorgeous mountains and picturesque villages with vineyards and apple orchards. We observed cattle and sheep grazing in green pastures. A shepherd stood leisurely tending his flock while his dog rested at his feet. Occasionally the lights would go out as the train roared right through a mountain from one valley to the next. Lakes and rivers of white water rushed on their journey to the sea.

In the dining car, we enjoyed a hearty repast of German wiener schnitzel, then returned to our coach seat and soon fell fast asleep. Shortly after we awoke, our train stopped and we lugged our carry-on baggage to another waiting train.

After a few preliminary jerks, we ascended the steep grade to Zermatt. The cog wheels of the train engaged with a loud bang on this always delightful ride. Our train chugged up to the next narrow valley and the next tunnel and the next bang of the cog wheels. After a few whistle stops we arrived at the end of the line—the little town of Zermatt, one of our favorite vacation spots. No automobiles or trucks were allowed. All transportation was by little battery-operated vehicles. The motors were used as generators, so going down one hill they would barely store enough energy to reach the top of the next. In the late afternoon people would have to scurry off the street into the nearest store or risk becoming entrapped among hundreds of sheep being driven to another pasture.

Once we retrieved our baggage, we located the porter and proceeded to the Corona Hotel on his electric cart. A beautiful little hotel, the Corona contained only a handful of rooms. Our customary room, the corner room on the second and top floor, had never appeared more inviting. The front windows looked out on the snow-capped Matterhorn, while the side window faced a fast-flowing river. Just across the river a cemetery held the remains of those who had ended this life during their last Matterhorn climb. Beyond the cemetery stood the Lutheran church whose tower bell helped us keep the time of day as we re-explored mountain hiking trails.

Our two-week vacation seemed to last only ten minutes. Upon our return to the Zurich train station, we took a taxi to the airport. I checked us in on the flight to Frankfurt. When the ticket agent recognized my name, she exclaimed in her limited English, "Go to phone—call this number—our police and your FBI looking everywhere to find you."

I rushed to the nearest phone. While in Zermatt we had purposely shied away from radios and newspapers to achieve complete relaxation. There was no television. What had happened? Had an explosion causing radiation exposure occurred at my laboratory, or were my services needed for a radiation accident in some other part of the world?

Floyd Culler, the associate director of ORNL, answered my phone call. He sighed with relief as he said, "Thank goodness, Karl, we've finally reached you. After you left Oak Ridge we reviewed your Neuherberg paper once again. Some changes must be made. I phoned Dr. Wachsman, the chairman of the meeting in Germany, and told him I was sending 250 revised copies. I told him to destroy the copies you had sent."

Furious but speechless, I restrained my anger.

Culler explained that he and laboratory management disliked my criticism of the LMFBR in favor of the MSTB.

Why this sudden shift of opinion? Only two weeks before, scientists at Oak Ridge had trumpeted the MSTB as *the* future power reactor. Some even speculated that it would make electrical power so cheap we would be able to do away with electric meters. Cheaper to operate than the PWR or

BWR power reactors, it did not require the very costly gaseous diffusion separation and concentration of the U-235 for the fuel elements.[6] The MSTB required only the relatively cheap chemical separation of U-233 from thorium.

We believed the MSTB and our unified system would be substantially safer than the LMFBR or the proposed BWRs or PWRs. The MSTB would never have a large inventory of fission products, because they would be removed constantly and prepared for shipment to salt mines for permanent disposal. The LMFBR, as well as the PWR and the BWR, would allow billions of curies of fission products to remain in the reactor for the fuel lifetime, essentially waiting for an explosion (like those that later occurred at Three Mile Island and Chernobyl) to smear them around the world.

Culler revealed the real reason management wanted to muzzle me when he admitted that I was getting crosswise in Washington with powerful elements in the nuclear industrial complex. The LMFBR had reached top priority in Washington (and ORNL coveted large contracts for its development). He told me, in effect, "Don't say anything about the superiority of the MSTB over the LMFBR. Don't you realize the president has decided to allocate $30 million of extra money to expedite building a demonstration LMFBR? You are jeopardizing the welfare of the laboratory."

Culler implied that if I were to present the original speech, hundreds of Oak Ridge jobs would be lost. Dumbfounded and flabbergasted, I slammed down the phone.[7]

Minutes later, Helen and I boarded the plane. At Frankfurt, a car met us and took us to Neuherberg.

Here, I made the biggest mistake of my life. I reasoned that if I fought the issue and hundreds of people in Oak Ridge lost their jobs, I would be one of them—I would lose not only my job, but also the retirement benefits I had labored over a quarter of a century to obtain. I feared that powerful elements within ORNL management would destroy my reputation in the scientific community. Even if I were to deliver my original speech, I rationalized, the nuclear industry would bury my warnings in a barrage of support for the LMFBR and keep the money box (the AEC) happy.

Red-faced, I bowed my head and described the risks of plutonium exposure, but without mentioning the MSTB or the LMFBR.[8] When I returned to ORNL, my fellow employees, disgusted with management, deplored this incident. W. S. Snyder, my assistant director, said it constituted censorship (fig. 37). Snyder was right. I should have stood my ground regardless of the consequences. Had I done so, perhaps the world would never have had reactors like those at Chernobyl and Three Mile Island. It is also likely that I would have returned to teaching and cosmic-ray research and forgotten all about health physics. In fact, I had already been working with members of the U.S. space program, warning about cosmic-ray dangers of supersonic commercial flight and orbiting space stations.

Ironically, nearly a quarter of a century later I learned that during the same period when I was making my greatest mistake, the laboratory director, Al Weinberg, was in the process of avoiding what would have been his greatest mistake. Although Weinberg supported Culler in the censorship of my Neuherberg paper, he never gave up his support for the MTSB to an extent that would satisfy the AEC and its subcontractor, Union Carbide Corporation.[9] In his excellent book *The First Nuclear Era*, Weinberg relates a discussion he had in 1972 with Rep. Chet Holifield, which left him "speechless" after Holifield warned him, "Alvin, if you are concerned about the safety of reactors, then I think it may be time for you to leave nuclear energy." Weinberg succinctly states, "I had never been fired before."[10] I retired in 1972, the same year that Weinberg was replaced

Fig. 37. Members of my team at ORNL during the early 1950s: (left to right) Mary Jane Cook, Walter Snyder, Mary Rose Ford, and Elda Anderson with Rolf Sievert (second from right) and me. Photo: ORNL)

as ORNL's director. After our respective departures, the health physics division was broken up and redistributed among other divisions. Effectively, it no longer existed.

A few months after the Neuherberg incident, laboratory management anxiously awaited my departure. When some senior employees at ORNL reached the required retirement age of sixty-five, they received the "privilege" of continuing at the lab on a one-dollar-a-year basis. This privilege was not extended to me.[11]

On September 27, 1972, I reached retirement age. The lab's parting gift to me, a clock run by changes in barometric pressure, constantly reminded me of what had happened at Neuherberg. Sometimes it seemed to say, "I am run by changes in air pressure and am here to remind you of your increase in blood pressure some years ago when you prayed quietly and asked God to help you bury the LMFBR lest it bury civilization."[12]

Early in October 1972 Helen and I moved from Oak Ridge to Atlanta. I refused a government job in Washington and accepted a professorship at Georgia Institute of Technology. I felt relieved to be teaching again and doing research as a member of the faculty of the School of Nuclear Engineering and Health Physics. By now rusty on some of my physics, I soon learned that I needed to study as much as my students.

During the nine years I taught at Georgia Tech, I developed a course on various types of radiation, including nonionizing, ultraviolet, infrared, microwave, ultrasonic, and infrasonic. I authored a chapter in a book on this subject.[13] The research projects of my graduate students interested and excited me. I served as thesis adviser for thirteen doctoral students.[14]

Research confirmed my earlier publications and long-held belief that plutonium is far more dangerous than admitted by the nuclear establishment.[15] One project we developed was a new method of dosimetry that had been previously researched by two visiting scientists under my direction at ORNL.[16] Our research with this new and more sensitive dosimetry again proved that the current maximum permissible concentration (MPC) for Pu-239 was more than 200 times too high.[17]

Several projects with my students G. B. Stillwagon and M. Sohrabi led to the development of a new type of fast neutron dosimeter free of the serious flaws of the proton track dosimeter in common use today.[18] One of our research projects at Georgia Tech involved using this dosimeter to determine a better value of the maximum permissible body burden of plutonium. Strong evidence that the maximum permissible body burden and the MPC values of plutonium are too high was also provided by R. P. Larsen and R. D. Oldham.[19]

Toward the close of my stay at Georgia Tech, I actively opposed work on the LMFBR at Oak Ridge. In 1982 I testified before a Nuclear Regulatory Commission panel regarding the flaws of the LMFBR. Tom Cochran, John Cobb, and I offered opponent testimony. Cochran had been a student of mine when I taught courses as an adjunct professor at Vanderbilt University. Both he and Cobb later become members of the Three Mile Island Public Health Fund Committee (TMIPHFC), which I chaired for a decade.[20]

On scientific merits, the three of us won hands down in the Oak Ridge hearing. We lost, however, as a result of obfuscatory reasoning on the part of the panel members and their desire to function as the NRC intended when it appointed them by the Tompkins Method.[21] Some senior members of ORNL bitterly resented my participation in this hearing in opposition to the LMFBR. They considered the LMFBR the future of ORNL operations since the laboratory had already spent millions on the initial phases of this program.

When I presented an invited lecture in Oak Ridge and again stated why the LMFBR should not be built, senior ORNL members viewed my address as the last straw. One of the laboratory directors of the LMFBR program wrote to Dr. Few, president of Georgia Tech, deploring my stand and shaming the university for having me on its faculty.

The next day the director of the School of Nuclear Engineering and Health Physics sent me a letter stating that my faculty appointment would not be renewed the next quarter because I had already passed the university's retirement age. A few days later I attended my retirement party. Georgia Tech's refusal to maintain my employment benefitted the university financially, since I had been a faculty member for only nine and one-half years. In order to qualify for retirement benefits, I would have to teach for ten years.

In retrospect, my termination proved a blessing in disguise. Now I could spend full time fighting for the people and principles most important to me. I could actively oppose not only the LMFBR but other bad choices of the U.S. Department of Energy.

CHAPTER FIVE

Nuclear Incidents in Other Facilities

The plants are safe: it's the people who aren't.

JOHN KEMENY, CHAIR OF THE COMMISSION THAT INVESTIGATED THE THREE MILE ISLAND INCIDENT

When the British scientist Sir John Cockcroft visited ORNL in the 1950s, he became convinced that the plutonium-producing plant under construction at Windscale, England, needed a filter system for the air from the reactor before it reached the discharge stack—a filter system similar to the one we had designed at ORNL.

Unlike the water-cooled, graphite-moderated reactors (WGRs) at Hanford, the reactors at Oak Ridge and Windscale were air-cooled, graphite-moderated reactors (AGRs). In retrospect, we can see that AGRs are very dangerous, not just because of the atmospheric releases of radioactive dusts and gases, but also because of their storage of Wigner energy in the graphite and the high exposures to the reactor operators. Fortunately—thanks to luck and to the work of the engineers and operators at ORNL—we managed to operate our AGR without a major accident!

At Windscale as well as our Oak Ridge operations, we knew from publications of Eugene Wigner that graphite stores Wigner energy after exposure to ionizing radiation. The electrons and electron holes are raised to a number of high energy levels in the graphite.

Any crystalline substance—diamond, salt, or graphite—stores energy when exposed to ionizing radiation. This was less of a problem for the

Hanford WGRs, because in those reactors the graphite and uranium slugs were constantly in a fast moving stream of water that carried off the heat as it was released—except, of course, when there was a blocked channel.

The situation was very different for the AGRs at Oak Ridge and Windscale. In these reactors the poor heat transfer from graphite to air allowed the storage of large amounts of energy in the graphite—energy that could be suddenly released without warning and could raise the temperature of the graphite and uranium slugs to a white heat before emergency rescue operations could be set in motion. The electrons and electron holes could be released without warning, resulting in a tremendous burst of energy, causing the temperature of the reactor to get out of hand. When the temperature was raised slowly—as was the weekly procedure at ORNL—the electrons and electron holes were released first at the lower energy levels and then at higher and higher energy levels so that the heat could be slowly dissipated. As the temperature continued to increase as a result of gradual reduction of the rate of air flow, energy was freed slowly from higher and higher energy levels in the graphite. This resulted in a tolerable temperature increase in the graphite and the aluminum-encased uranium slugs. Still, the procedure had to be done gingerly and with extreme caution. It was like slowly opening the gate in a dike to release water pressure lest the dike give way.

This reaction, called the Wigner effect, was named after ORNL's Eugene Wigner, only one of his many significant contributions to science. Wigner received the Nobel Prize in physics in 1963. He should have received a second Nobel Prize for his discovery of the Wigner effect, which prevented several reactor accidents, but he did not.

The Wigner effect significantly contributed to accidents at Windscale and also at Chernobyl and Three Mile Island. At Windscale, as at ORNL, the Wigner energy in the graphite was released slowly at frequent intervals. But on October 10, 1957, as technicians were releasing the Wigner energy from the Windscale reactor by restricting air flow, the temperature in portions of the reactor (which was much larger than the one at ORNL) rose too rapidly. This melted the aluminum cans around the uranium slugs and set the graphite and uranium on fire.

Again, as had happened in our Oak Ridge reactor accident, the engineers were afraid to spray water into the reactor for fear of an explosion from an aluminum-water or uranium-water reaction. Finally, they had no choice but to spray a large amount of water into the reactor to quench the fire. Fortunately, they extinguished the fire without an explosion. Nonetheless, this constituted a serious accident.

Soon afterward I received a phone call from my friend Greg Marley, a leading British health physicist. He coupled his frightening message concerning the serious accident at the Windscale reactor with a plea for help. A couple of days later I arrived, having flown to London and taken a train from the London airport, at the Windscale plant, located in a little town on the Irish Sea.

The English were waging a surreptitious battle against the Irish as well as the Welsh when they dumped radioactive waste into the Irish Sea. Fission products and plutonium that came in contact with the oysters, fish, and seaweed that the Irish and Welsh ate contaminated this portion of the food chain. Despite our knowledge of the hazards of radioactive waste and complaints from many sources, the Windscale plant continued for years to dump its waste into the Irish Sea. Some of it washed up on the beaches where children played and was tracked into homes.

The Windscale health physicists shocked me by casually noting that there was a considerable amount of bismuth in the reactor at the time of the fire.[1] The fuming bismuth formed a fog of relatively large particulates on which iodine-131, strontium-89 and Sr-90, and cesium-134 and Cs-137 occluded. The filters in the stack filter house, wet from water used to put out the fire, held up the large bismuth particles with fission products on their surfaces.[2] The English considered this a piece of good luck, since normally wet filters are inefficient.

Did I correctly hear them when they said they had bismuth in their reactor? Every document I had received about bismuth in reactors over the past few years was in a red cover marked "Top Secret!" A special courier would deliver these documents and literally stand guard in my office until I had locked them in my safe. I had been led to believe one might receive the death penalty for discussing the use of bismuth with anyone who did not hold appropriate security clearance.

Although Cockcroft's Folly greatly reduced the harmful discharge into the environment, radioactive contamination many thousand times that released later in the Three Mile Island accident escaped that day into the English countryside. So far as I was able to discern, no alpha measurements (those that would detect Po-210) were made in the Windscale environment. Polonium-210 is an intense alpha emitter, which the ICRP considers thirty times more harmful than gamma radiation once it is deposited in the human body.[3]

In retrospect, even though we could only discuss Po-210 with those who had top-secret clearance, I should have shouted my concern. It was absurd to retain Po-210 on the top-secret list as late as 1957, some eight years after the Russians had exploded their first atomic bomb. But I never violated the security clearance regarding Po-210 and returned to Oak Ridge somewhat red-faced. Now, some forty years later, time and the law of entropy have erased most evidence of the Po-210.[4] But undoubtedly, during the first few months after the accident, Po-210 presented by far the most serious effects. Many people who were downwind from Windscale probably developed cancer from the Po-210 released in the accident.

In most other respects the English team performed well trying to contain and minimize the consequences of the accident. With their Geiger counters they visited farmers in the outlying areas. They would hold their Geiger counters against the sides of a typical stainless-steel 5-gallon jug of milk. If the reading registered too high, they would pour out the milk. They discouraged consumption of garden produce and recommended that children take baths after playing outdoors.

The tragedy was that they failed to check for alpha radiation activity. Geiger counters measure only beta and gamma activity. Even at that early date, alpha dose in equivalent roentgens was considered ten times more harmful than the dose from beta or gamma radiation. Today the ICRP considers alpha doses thirty times more damaging to human tissue than beta or gamma radiation. In 1957 proportional counters were certainly available to measure alpha activity, but I saw none in use at Windscale.

The British scientists considered it unfortunate that their light-aircraft survey equipment did not become airborne until three days after the accident. I carried this information back to ORNL, and we initiated a program that would enable our planes to be airborne in less than an hour following a serious accident.

The English did not have an outside facility to assemble exposure data and pass the information on to the news media. We learned from this as well and arranged to have such a facility available for collection of data and for communications in case of such an accident. Years later, the operators of the Three Mile Island nuclear power plant never examined the pages of history that described the Windscale accident and did not take the precautions that were implemented at Oak Ridge. They never did get light aircraft airborne at Three Mile Island. A long time later they sent up a helicopter. In some respects this was worse than nothing, because the air turbulence generated by the helicopter blew away and partly wiped out what they were trying to measure. They also did not realize that they should be looking for the alpha emitters. Lacking a facility for assembling data on radiation measurements and passing it on quickly to the news media, they invited many false and scary media reports.

A nuclear accident in Yugoslavia on October 15, 1958, also caused great concern. Our health physics group was invited to Vinca, Yugoslavia, under the leadership of G. S. Hurst, to perform the dosimetry of the accident. Before the accident workers were calmly performing their duties not far from the reactor—while an annoying Geiger counter nearby clicked away at a very high rate.

For some time this Geiger counter had been malfunctioning and had given many false alarms, so no one paid any attention to it. The reactor operator, while keeping an occasional eye on the reactor control panel, quietly made the "best use" of her time by studying to improve her knowledge of English. Fortunately, someone entered the room carrying a Geiger counter that was functioning properly, and it began to click wildly. At that instant, everyone in the room realized they were in an intense radiation field, and they ran from the building—the only effective safety measure in such a situation.

The reactor operator received a dose of 436 rem, according to Hurst's best estimates.[5] Although she was given extensive bone marrow treatment, she died thirty-two days later. The treatment she received in a Paris hospital was given by the so-called push-pull method, in which the donor lay on a cot beside the patient while a sample of active (red) bone marrow was drawn from the sternum of the donor and immediately injected into the bone marrow of the patient.

The Idaho Falls SL-1 reactor accident on January 3, 1961, also proved to be tragic. The U.S. Navy was just starting up this small experimental reactor. The chain on top of the reactor that pulled the main control rod up out of the reactor had a history of binding. A young officer climbed to the top of the reactor, leaned over, and pulled on the top of the control rod as he had often done before to keep it from binding. But this time it came up too fast, making the reactor prompt critical. A terrifying explosion occurred, in which the top of the reactor blew off and the reactor control rods blasted skyward like shells from a powerful howitzer.

One of the control rods pinned the officer on the ceiling. Two other men were blown to the far side of the building. The SL-1 area alarms screamed. When the first health physicists arrived a few minutes later, one of them ran up the stairway of the building, where he saw the horrible sight of the man pinned on the ceiling. Also he observed another man in the far corner of the building who appeared to move his arm. The health physicist set his Geiger counter on the highest scale, 10 R per hour, but the counter was paralyzed, the pulses coming so fast that they could not be resolved, so the counter failed to click and the instrument read zero.[6]

Realizing that a paralyzed Geiger counter can mean a dose rate of thousands of R per hour, he rushed back down the stairs and away from the building lest he, too, should become one of the fatalities.

As soon as he realized the seriousness of the SL-1 accident, John Horan, the senior health physicist at the Idaho Falls operations, telephoned to ask me to hurry to Idaho Falls. Horan had been one of my early health physics students at Clinton Laboratories.

Upon arrival I learned that one of the bodies had been recovered. Rescue workers surrounded the corpse with 2-inch-thick lead bricks. I insisted that autopsies be performed on all three bodies, but Horan, the AEC, and Idaho Falls officials remained reluctant even after I argued with them for hours into the night.

Determined to conduct my investigation properly, I called Dr. Clarence Lushbaugh, a pathologist at Los Alamos. Lushbaugh immediately grasped the gravity of the situation and said, "I'm on my way to the airport and will be there as fast as the plane can fly." There was a long delay before the other two bodies could be recovered because of the deadly high radiation levels in the building.

Lead sheets were placed about the bodies during the autopsy in order to protect Dr. Lushbaugh. Lead was even placed in the coffins during the funerals to offer protection until the bodies could be buried under 8 feet of soil.

The autopsies revealed the cause of the accident and the dose to each body organ of the three victims. Blood and hair samples allowed us to measure the dose. Hair samples from various parts of the body of the man pinned on the ceiling by the control rod were particularly revealing. These samples, which we measured in our own laboratory, indicated a fast, high-energy neutron dose had reacted with sulfur in the hair, producing P-32. This is how we knew he had been stooping over while pulling on the control rod when it was ejected. Sodium-24, produced by thermal neutron induction from sodium-23 in the blood, provided useful information on the integrated, total body dose.[7]

In many respects the SL-1 reactor operation in Idaho was an accident waiting to happen. Since radiation cannot be detected by human senses, it is easy to understand why persons working with it develop a false sense of security. The "shoestring" operation at Idaho Falls resulted in an extremely dangerous experimental device. The main control rod remained stuck for several months. Rather than shut down the operation and repair the reactor, a man was sent to the top of the reactor to jerk the hoist chain on the main control rod.

So-called "poisons" in the reactor had been burned up, causing an inadequate safety margin on the control rods.[8] Thus the reactor operated with too many control rods inserted, and rather than shut it down for a long-overdue major overhaul, unskilled operators simply pasted strips of cadmium along the walls of the reactor vessel. These strips acted as shims and held down the activity of the reactor to the proper operating range.[9] After awhile these cadmium strips fell off the sides of the reactor vessel, causing the reactor to "burp" or suddenly increase power. Disregarding these burps was wildly irresponsible in the eyes of a health physicist.

The three men who died in this accident made a great sacrifice, and perhaps it was not all in vain. Their recklessness and this serious accident with an experimental reactor served as a lesson of vital importance to Admiral Rickover of the U.S. Navy. Perhaps it had the same effect on Rickover as the mistake made by Whitaker years before when he disregarded health physics safety rules of our operations at ORNL.

The Chernobyl nuclear power station stands on the bank of the Pripyat River, which flows into the Dnieper. On April 26, 1986, at 1:23 A.M., the operators of Chernobyl decided to use the shutdown time for annual maintenance as an opportunity to test how long the power station's steam turbines could produce electricity after the steam source had been cut off. But, as one chronicler of the Chernobyl disaster has written, "the technicians who drew up the plans did not discuss them with the physicists or other nuclear safety staff. . . . And from the beginning, the operators seemed hell-bent on self-destruction."[10]

An early, critical mistake was to shut off the emergency core cooling system, an act that deprived the reactor of one of its vital safety systems.[11] As a result of an incorrect regulator setting, a not-infrequent operator error in many plants, the reactor's power sank to 30 MW instead of stabilizing between 700 and 1000 MW. Another error was to unknowingly alter the balance of steam and water in the circuit, thereby making the reactor extremely unstable.

Next, in response to sagging steam pressure and a water level that dropped below the emergency mark, the technicians blocked the automatic shutdown system that normally would have closed down the reactor.[12] For unknown reasons, perhaps panic, they switched off the last safety system. At this point the reactor ran out of control. "In the last second of the reactor's life its power surged from 7 percent to several hundred times its normal level. A small part of the reactor's core went 'prompt critical.' The effect was the equivalent of half a ton of TNT exploding in the core. In fact, it was very much like the detonation of a small atomic bomb—something that the nuclear industry had always insisted could never happen in a reactor."[13]

Only the discovery in Sweden of substantially increased levels of radioactivity and Swedish diplomatic inquiries finally prompted the Soviet Union to issue a terse statement admitting that an accident had occurred. By the time of the Soviet announcement, radioactivity had traveled across both eastern and western Europe.[14]

Not until five days after the disaster was the town of Chernobyl evacuated. On that day dramatic farewells took place between masters and pets, since a group of hunters was required to shoot all the dogs in the town.[15]

One of the victims of Chernobyl was Vladimir Pravik, a fire fighter. Pravik stoically endured the pain of the acute radiation syndrome as a result of trying to put out the fire. "Pravik had no skin at all, having lost it to radiation, which also destroyed his salivary glands and left him with a mouth as dry as soil in the midst of a drought. That was why he still could not talk. He just looked and blinked with his eyelids, which no longer had any eyelashes. . . . He began to wither and dry up with the approach of death, as the skin and body tissues of radiation victims do in fact become mummified. In this nuclear age, even death is transformed and made somehow less human, as the dead are blackened, shriveled mummies, as light as a child."[16]

For those thousands and tens of thousands whose names I do not know, now and in years to come, I am certain they too will die from cancer, suffer anemia, undergo genetic defects for generations, battle hypothyroidism, or endure mental retardation as a direct result of Chernobyl's radioactive fallout. Even so, the disaster was not as bad as it could have been, thanks to two fortuitous circumstances. First, a slight wind carried the fallout primarily over sparsely populated forest. Second, the disaster occurred at night, when few people were out. The vast majority of the population were indoors and therefore shielded from 90 percent of the fallout.[17]

Despite numerous incidents inside nuclear facilities, influential sources within the industry downplayed them. It may be difficult for some to appreciate this unyielding stance. Money and its frequent companion, greed, offer a partial explanation. Once a cosmic-ray physicist and then a nuclear zealot, I understand that their obsession also has noneconomic links. We had studied the stars (nuclear power plants themselves), and now we could turn on "little suns" at will to heat or cool our homes or change deserts into gardens.

I can attest from long experience that anyone who challenges nuclear power must be prepared to withstand political, economic, and professional attacks. For example, when I publicly criticized the vast majority of health physicists (for not stepping forward to assist injured workers in cases) during a keynote speech in 1985 before union workers, Dr. Clarence Lushbaugh promptly responded in the *Oak Ridger* by equating me with the lowest species of "animals that befoul their own nest."

If we dare to uncover our eyes and pause a few moments for thought, we must realize that the world's reactors are not acceptably safe. Where do we hide nuclear wastes, including thousands of radioactive waste tanks threatening to leak and possibly explode as one did in the former Soviet Union? What do we do about the hundreds of thousands of curies of radioactive noble gases and tritium (Xe-133, Xe-135, Kr-85, and H-3) released annually into the air we breathe? Until we understand the price to humankind, we will be unable to develop acceptable solutions.

CHAPTER SIX

The Price

We have found the enemy and he is us.

WALT KELLY'S "POGO"

RELUCTANT AEC

From the beginning of operations at Oak Ridge in 1943, we recognized the existence of a major unsolved problem: how to dispose of radioactive waste. The work at Y-12 involving electromagnetic separation of uranium-235 from natural uranium and weapons development, the work at K-25 on gaseous diffusion separation of uranium-235, and our own operations at X-10 exploring the separation of plutonium from uranium and fission products: all involved getting rid of a massive mixture of chemical and radioactive wastes. The operations at Y-12 and K-25 resulted in radioactive waste of uranium and thorium and their natural radioactive daughter products, as well as some manmade radionuclides. The most serious waste problem at these facilities, however, concerned chemical wastes, such as mercury, lead, arsenic, cobalt, uranium, thorium, lithium, and tin. Our X-10 facility served as a radioactive waste cauldron for over 500 different radionuclides with half-lives ranging from fractions of seconds to millions of years. Some of these radionuclides, such as plutonium, were among the most hazardous materials known.

When I initially sought research funding to study the disposal of radioactive waste and radiation ecology, our funding source, the Atomic

Energy Commission (AEC), was unwilling to recognize the urgent need for this research. As early as the late 1940s we requested $2 million to explore methods of dealing with the radioactive waste and ways of protecting the environment.

Orlando Park, a distinguished ecology professor at Northwestern University, E. G. Struxness, my assistant director, and I developed a broad plan (fig. 38). We told the AEC that it had a responsibility to protect arthropods, bacteria, fungi, trees, and animals exposed to our radioactive wastes. The committee did not take us seriously. Even once the AEC realized how determined we were, one official commented, "Man is the thing we should be interested in protecting; we should protect him and forget about these microorganisms and other forms of life. After all it would be a good thing if radiation destroyed all these microorganisms." We were shocked by this response. Didn't they realize humankind could not survive without some of these organisms in our bodies and in the environment?

Another response to our proposals was: "Why not just dilute the radioactive waste to the occupational maximum permissible concentration, discharge it into White Oak Creek where it will seep into the Clinch River, and forget it?" I reminded the official in question of our studies, which indicated that certain radionuclides concentrate in some plants and

Fig. 38. Ed Struxness (second from right), my assistant director, who helped me prepare a broad funding plan for research on radioactive waste disposal and radiation ecology, with (from left) Sam Shoup, Elda Anderson, Fred Cowan, Walter Jordan, and me. (Photo: ORNL)

animals by factors of 100 to as high as 10,000. No sooner had I responded to this environmentally insensitive bureaucrat when another AEC official likewise objected to our funding request, saying, "Man is the most radiosensitive element in the ecosystem. Protect him and forget about microorganisms, many of which take 100,000 R to destroy." I responded with strong skepticism—which was vindicated some years later when research showed that pine trees are destroyed at about 100 rem, while the mid-lethal dose for a human being is between 150 and 400 rem.[1]

Initially our only feasible option was to store the uranium waste in underground holding tanks. We also provided short-term retention in holding ponds for millions of gallons per year of liquid waste containing hundreds of thousands of curies of mixed fission products.

The ponds served a useful purpose for short-lived radionuclides since they provided good waste hold-up until these radionuclides decayed into stable elements. Unfortunately, some of the ponds developed bad leaks. Findings from the test wells that we dug showed that the patterns of flow from these impoundments did indeed tend in the direction of White Oak Lake, which in turn contaminated the public domain via the Clinch River.

The waste was accumulating so quickly that the hold-up time became very short. So we renamed the holding ponds "settling basins." In retrospect, I should have put my job on the line and insisted that operations be curtailed until we discovered a better solution.

Initially I reasoned that winning the war was so important that the unknowns of how to dispose of radioactive waste should not impede our programs. During the Cold War we developed two very promising methods for disposal of high-level radioactive waste: by storing it in salt and by hydrofracturing shale formations. Unfortunately, these studies were discontinued when I departed ORNL in September 1972. Today issues of radioactive waste disposal contribute to the stalemate within the nuclear industry. Al Weinberg, our former laboratory director, in his book *The First Nuclear Era* dedicated less than one page to waste disposal at ORNL. He succinctly stated, "Instead of waste disposal becoming a central focus for the Laboratory, the matter was always a side issue—one that never commanded the attention of the most sophisticated people. I cannot give a complete explanation of this, except to say that the ad hoc waste disposal methods used at Clinton seemed adequate at the time."[2]

During the 1950s the problem of disposal of radioactive liquid waste reached alarming proportions. From June 1952 through 1959 we disposed of more than 15 million gallons of contaminated waste. By 1960 even the AEC listened to our concerns. We received generous financial support for our research, but we were never able to catch up with the problem. What we did accomplish was abruptly halted when I left ORNL.

Rivers of Radioactive Waste

The contamination of U.S. waterways with radioisotopes began in facilities such as Oak Ridge. Our initial ignorance is understandable, but I am most disappointed that fifty years later we have made little advancement in the disposal of radioactive waste, even as quantities of radioactive waste have increased exponentially throughout the world. Understandably, no one wants radioactive waste in his backyard.

In 1943 none of us knew about the long-range genetic, teratogenic, and somatic risks of low-level exposure.[3] But in spite of our ignorance of the enormous health risks, we were acutely aware that the Oak Ridge concentration of curies of radionuclides was millions of times greater than ever experienced before in human history. The pile building caused us at times to be apprehensive about our ability to safeguard a working environment. If it had not been for the 6 feet of concrete shielding, the levels of gamma radiation at this building would have been millions of times what we thought was the mid-lethal dose—if human beings were as radiosensitive as mice and guinea pigs.[4]

Neutron exposure from the pile building concerned us because there were no data except our calculations to show that it might be harmful. A British scientist, J. Chadwick, had discovered the neutron only eleven years earlier, in 1932.[5] Now we had millions of curies of fission products, comprising over a thousand different radioisotopes with half-lives ranging from fractions of a second to millions of years—radioactive waste looking for a place to hide.

Our first measure was to construct a small earthen dam across White Oak Creek a quarter of a mile above its out fall to the Clinch River. This dam formed White Oak Lake, located two miles from the laboratory, with a capacity of 10 million cubic feet.

Under normal flow conditions, with a closed gate, White Oak Lake provided a twenty-three-day delay in release to the Clinch River. In the early period, we kept the gate closed. Water samples were collected several times a day and analyzed for radioactivity.

The Clinch River partly surrounds the Oak Ridge reservation. Our riskiest practice related to disposal methods for liquids. ORNL drained radioactive waste almost entirely into the White Oak Creek and Lake system, which emptied into the Clinch River that flowed past the Y-12 and

K-25 areas (figs. 39, 40). From here some of these radionuclides found their way down the Tennessee (about twenty miles downstream from White Oak Creek), Ohio, and Mississippi Rivers, eventually ending up in the Gulf of Mexico.

In November 1960 we commenced a radioactivity study in the Clinch and Tennessee Rivers. We established sampling stations along these waterways, so that numerous samples could be collected no less frequently than every week. We then analyzed the samples to determine whether radionuclides were present. The most common radionuclides we found in this

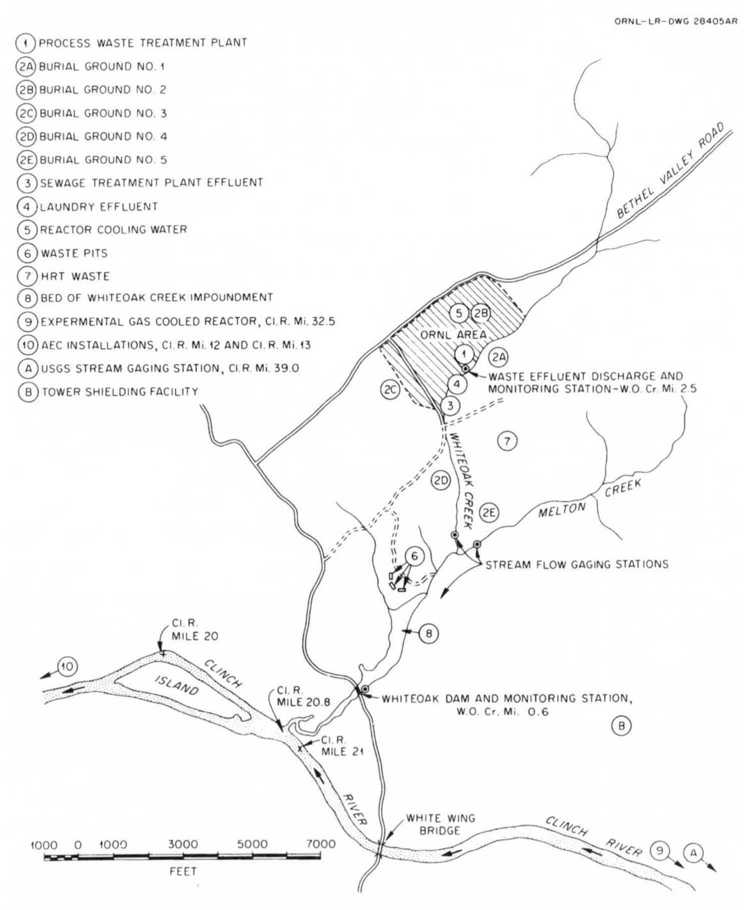

Fig. 39. Map showing sources of radioactive contamination in the ORNL environment. (Map courtesy of ORNL)

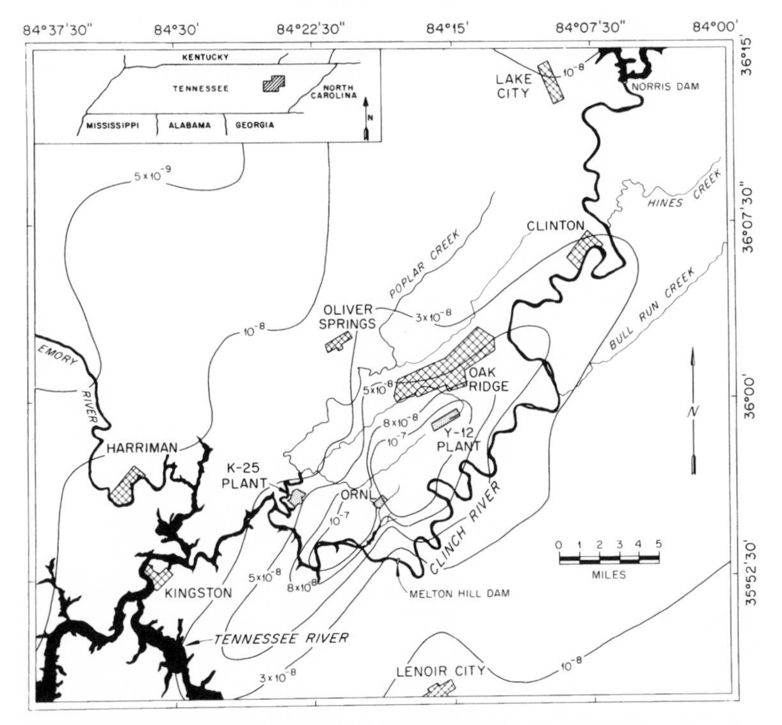

Fig. 40. Map showing the path of waterway contamination in the region of Oak Ridge. Once radionuclides from the White Oak Creek and Lake system emptied into the Clinch River, which flows by Y-12 and K-25, they found their way to the Tennessee River and from there to the Ohio and Mississippi Rivers, eventually ending up in the Gulf of Mexico.

river water were cesium-137, cobalt-60, ruthenium-106, and strontium-89 and 90. It was disturbing to learn that a significantly high percentage of ruthenium released to the Clinch River from White Oak Creek was found in samples collected at Chickamauga Dam on the Tennessee River, some 120 miles downstream.

Watts Bar Lake, an impoundment of the Tennessee Valley Authority on the Clinch River, at crest would back up to the out fall of White Oak Lake and empty into the Tennessee River, whose waters eventually flowed into the Gulf of Mexico via the Ohio and Mississippi Rivers. The silt just off Watts Bar Lake was severely contaminated not only with radionuclides but also with mercury, lead, and other chemical poisons from the Y-12 plant. In this case the chemical waste entered the Clinch River via Poplar Creek.

Watts Bar Lake, as well as the Clinch River, provided drinking water for several cities, despite the presence of official signs warning persons not to swim in or drink the water. The radioactive contamination that emptied into the Tennessee River from Watts Bar consisted of Sr-90, Ru-106, Ce-144, Cs-137, Co-60, and Zr-95.

Great variation occurred over the years in the amount and composition of the radionuclides that left White Oak Lake and discharged into the public domain via the Clinch River (see table). These variations stemmed in large part from changes in chemical separation techniques.

During my time at ORNL we monitored White Oak Lake and environs daily, the Clinch River weekly, and the Tennessee River monthly. There are many villages, towns, and cities along these rivers—and many more beyond the Tennessee. We even made a few measurements in the Ohio and Mississippi Rivers. Radionuclides with long half-lives—such as I-129 (17 million years), C-14 (5,715 years), Np-237 (2.14 million years), Pu-239 (24,110 years) and Pu-240 (6,537 years)—had plenty of time to reach the Gulf. Unquestionably some of them did.

In 1944, the engineering division asked me to set the level for water in White Oak Lake, which impounded essentially all the liquid radioactive waste released from the laboratory, at 100 R per day.[6] We constructed ponds, trenches, and pits to retain this waste. They were of many types and sizes, and located in a variety of soils—clay, shale, loam, and sand. Some were lined. We added various absorbent materials and chemicals to the liquid waste. We experimented with a variety of pH values and other variables in the pits, including limestone gravel.

The requested limit would be double the limit set by the International X-Ray and Radium Protection Committee (IXRPC) in 1934.[7] Stubbornly I held out for a level 1000 times lower. The limit of 100 R/day would have been more than 9 million times the level of 0.004 mrem/year set for potable water by the U.S. Environmental Protection Agency in recent years. I deserve no credit for my stand, however, because even the level I set would have been over 9000 times the current EPA level. The engineers, for their part, argued that 100 R/day was reasonable since White Oak Lake would be fenced off and no one would be permitted to drink water directly from the lake or swim there. I feared that someone might swim in it or fish there. I was also concerned that there might be concentrating factors in the food web of which human beings are a part or that "hot fish" could escape into the public domain of the Clinch River.[8] For example, we soon discovered that birds rested and fed on these highly contaminated ponds, so we covered the ponds with chicken wire. Some of the contaminated

Yearly Discharges in Curies of Radionuclides into the Clinch River, 1949–61

Year	Gross Beta	Cs^{137}	Ru^{106}	Sr^{90}	Total Rare Earth Less Ce	Ce^{144}	Zr^{95}	Nb^{95}	I^{131}	Co^{60}
1949	718	77	110	150	77	18	180	22	77	
1950	191	19	23	38	30		15	42	19	
1951	101	20	18	29	11		4.5	2.2	18	
1952	214	9.9	15	72	26	23	19	18	20	
1953	304	6.4	26	130	110	6.7	7.6	3.6	2.1	
1954	384	22	11	140	160	24	14	9.2	3.5	
1955	437	63	31	93	150	85	5.2	5.7	7.0	6.6
1956	582	170	29	100	140	59	12	15	3.5	46
1957	397	89	60	83	110	13	23	7.1	1.2	4.8
1958	544	55	42	150	240	30	6.0	6.0	8.2	8.7
1959	937	76	520	60	94	48	27	30	0.5	77
1960	2190	31	1900	28	48	27	38	45	5.3	72
1961	2230	15	2000	22	24	4.2	20	70	3.7	31

Note: Values calculated from data supplied by Applied Health Physics Section. From ORNL "Health Physics Division Annual Progress Report for Period Ending June 30, 1963," 45.

ducks we banded were later retrieved from distant states and even as far away as Canada. (What about those unbanded contaminated ducks that in all probability were eaten by a hunter and his family? Did some of these innocent people later develop cancer?)

After about twelve years of laboratory operations, the lake reached a state of equilibrium where as much radioactive material left the lake as entered it. Our engineers informed us that we could expect a cloudburst of rain every twenty-five years that would wash out the dam. This, together with the fact that a large fish population had developed in the lake, prompted us to drain the lake and build a much better reinforced dam. Before draining the dam, we killed the fish and determined the concentration factor of the various radionuclides in bone, liver, and muscle. We found concentration factors as high as 10,000 for some of the radionuclides—a sobering result indeed.

The data we collected from the water and mud samples of White Oak Lake and the Clinch and Tennessee Rivers provided surprises. For example, we found high levels of radioisotopes (Cs-137, Sr-90, and Ru-106) in mud samples collected from the bottom mud several miles up the Emory River, above its confluence with the Clinch River (fig. 40) and not far from the water intake of the city of Harriman. Surely, I reasoned, water doesn't flow upstream.

After considerable deliberation I realized that the water in the Clinch was colder than that of the Emory. The density of water increases with dropping temperatures down to 4°C, so the cooler Clinch water would flow up the bottom of the Emory until it had warmed to the temperature of the Emory River water. In the process, much of the suspended mud with its complement of radioisotopes was deposited on the Emory River bottom. Part of the time water from the Clinch would back up the Emory when the Watts Bar Dam on the Clinch below Kingston was fully closed. This facilitated the precipitation of the mud particles in both the Emory and Clinch Rivers.

The predominant soil formation in the waste pit and pond areas, as well as at White Oak Lake and the Clinch River below White Oak Lake outfall, is Conasauga shale. Fortunately it showed a moderate to high retention factor for nearly all the fission products. Ruthenium-106, with a half -life of 372 days, was the exception. The good news is that only about 3 percent of the Ru-106 in the pits was released into White Oak Lake. The bad news was that a 3 percent release factor during 1957, 1958, and 1959 represented 200, 160, and 1300 curies, respectively! Very quickly this material found its way to the Clinch River.

The soil held up most of the uranium, plutonium, and transplutonic radionuclides present in low concentrations in the waste. It also held up most of the induced activity—cobalt-60 and phosphorus-32.

Some of the radionuclides, especially those with short half-lives, decayed into stable elements.[9]

A small fraction of the radionuclides produced were shipped for use in hospitals, universities, government and private research facilities, and industry. These were short-lived radionuclides, such as I-131 and P-32, which were used for conducting medical tests, such as injections of trace amounts before bone scans. The rest decayed radioactively or escaped into the environment, mostly via White Oak Lake and the Clinch, Tennessee, Ohio, and Mississippi Rivers. However, gaseous radioactive wastes with half-lives from hours to millions of years were also released, from the plant stacks. How much damage will be inflicted upon future generations as a result of these practices?

Even in the Clinch and Tennessee Rivers some of the radioactive contamination was not the fault of Oak Ridge. The high levels of cobalt-60 (5.271 y) that we found in the Clinch River in the mid-1950s perplexed us. Mysteriously, during one short period the levels of Co-60 in the Clinch exceeded those in White Oak Lake. Even more confusing, the Co-60 concentration increased in water samples taken a few miles up the Clinch above the White Oak Creek outfall. We ruled out backflow this far upstream. Then we asked ourselves, did this come from fallout from airborne contamination originating from our laboratory?

There was no record of any recent accident with Co-60, so we extended our river survey thirty miles upstream from the White Oak Creek outfall. By the time we reached fifteen miles upstream, we no longer had to collect water samples for our low-level analysis and counting facility at the laboratory. From here up the levels of radiation were so high we could take readings directly from our Geiger counters as we walked up the riverbank. While so engaged, one of our health physics surveyors yelled, "This creek is as hot as a firecracker!"

After advancing another quarter of a mile, we switched our counters to the high scale. Looking ahead we observed a flimsily constructed building with 55-gallon steal drums and piles of trash outside it. Near some of the drums our Geiger counters ran off-scale on the highest setting and blacked out. Levels of airborne contamination inside the building exceeded by 100 times what we considered "acceptable" at ORNL.

It turned out that the AEC was actually licensing this small, fly-by-night outfit to manufacture Co-60 and Cs-137 sources for medical and commercial use. The owners failed to provide the workers with any health physics protection and showed little or no concern for radiation exposure. The facility was shut down shortly after we reported our findings to the AEC.

Radioactive contamination of U.S. waterways extends far beyond the southeastern United States. H. M. Parker, director of the health physics program at the large plutonium-producing facility at Hanford, Washington, in 1948, told me that he found concentration factors as high as 1 million in bottom-feeding fish in the Columbia River. The Columbia supplied cooling water for the Hanford graphite-moderated, water-cooled reactors (GWRs). This great western water source contains minerals, such as natural sulfur, that become radioactive when exposed to a high flux of neutrons, such as that in the Hanford reactors.[10] The radioactivity of the Columbia came from phosphorus-32 (14.28 d) found in fishbones.

During the early operating period the Hanford facility passed the river water through the reactors with almost no pretreatment and allowed it to reenter the river without subsequent treatment or sufficient delay time. The use of delay ponds to allow decay of short-lived radionuclides would have been an advantage and should have been implemented from the beginning of operations. During this time period no one informed inhabitants downstream about the substantial health risks from drinking water or eating fish taken from the Columbia. The Hanford facility hurriedly began to take remedial measures while rushing to make enough plutonium for the bombs used at Alamogordo, New Mexico, and Nagasaki, Japan.

Even as late as 1961, the AEC received a report that concluded that the contamination of the river was at a crisis: "Thus, in terms of the resulting human radiation exposure, the capacity of the Columbia River to receive further radioactive pollution appears to be nearly, if not fully, exhausted for practical purposes, at least in the vicinity of Hanford Works." In its first ten years of operation, Hanford also had more than a half-million curies of airborne releases of I-131![11]

Today, the Hanford facility presents yet another environmental hazard. The weapons production plant at Hanford holds highly radioactive waste in 177 underground tanks. The tanks are either leaking or waiting to leak into the desert sands. Some of these radionuclides will be around for a thousand years. Will they ultimately flow to the nearby Columbia River? Or could there eventually be inadequate monitoring, leading to an explosion of one of the tanks (this actually happened in the former Soviet Union)?

Over the years Helen and I made several trips to Yugoslavia and Hungary, where I set up training programs in health physics. Often in these countries a meal would begin with fish soup. It contained chunks of fish—bones and flesh—and I did not find it very palatable, but to get the calcium I would eat mine, and to be polite I would pretend to like it. Of course, I knew that Sr-90 and Sr-89 can be concentrated by a factor of at least 10,000 in the skeletons of some fish. But even in my worst apprehensions, I never dreamed that the fish in my bowl might come from certain rivers or lakes in the Soviet Union where it would be highly contaminated.

The reports we received years later from some parts of the former Soviet Union confirmed extremely large exposures to Russians from eating fish soup. Although we might instinctively recoil in horror at the thought of the radioactive waste practices in Russia, we must never forget our own practices at Hanford.

FROM HUMAN GUINEA PIGS TO WEAPONS TESTS

Throughout the war period, Pentagon officials would visit me at ORNL to discuss the feasibility of using fission products as an adjunct to chemical warfare. I discussed this with Robert S. Stone, the associate director of health of the Manhattan Project. From 1943 to 1945, Stone's office was situated next door to mine in the wooden barracks at ORNL. This building served as headquarters for both health physics and the medical department in 1944.

At the biology, health physics, and medical seminars held every two months in one of the Manhattan Project sites—Oak Ridge, Argonne, Los Alamos, Hanford, Berkeley—we considered this question of a lethal fission-product weapon. I visited Dugway, New Mexico, the U.S. chemical warfare testing site. The specialists there agreed with me regarding the military utility of this weapon. It could be dropped on enemy factories with little damage to structures. By excluding from the weapon radionuclides with a half-life greater than eight days, the radiation level would drop to less than 1 percent of its original level in less than two months.

But one item of essential information was always lacking: data on how we might expect human beings to respond to various high levels of radiation exposure. As a result the other members of the ORNL Health Physics Division and I conducted cruel studies on monkeys and baboons. At doses of 5000 R they still responded, though sometimes weakly, to the proper routine, such as retrieving a banana dropped automatically into the cage if they pushed the correct buttons in proper sequence. However, from 5000 to 50,000 R food provided no incentive, so instead we gave the poor animals an electrical shock if they made mistakes.

More revolting than our treatment of animals was the concomitant disregard for humans. The release of 500,000 curies of I-131 (8.04 d) into the environment at Hanford was permitted in part to complement the chemical warfare studies of the effects on the neighboring population downwind—and, of course, away from the Hanford community. The rationale was the need to test their instruments in calibrating the fallout pattern. The exposed "expendables"—primarily Native Americans—were not told of this until 1990, some forty-five years after the exposure. Recent studies by one of my former Georgia Tech students, J. E. Till, revealed that doses to the thyroid reached hundreds of roentgens. Not surprisingly, a high incidence of thyroid carcinoma has appeared in this population.

Studies in some of our hospitals were even more revolting than the Hanford misdeeds. Hundreds of hospital patients were used as guinea pigs by their doctors—doctors whom the patients trusted to relieve their suffering and, if possible, restore their health. In numerous instances, no benefit resulted from the radiation treatment these patients received, ostensibly to combat their disease. Injections, pills, operations, and forced inhalation of radioactive substances hastened death, caused excruciating pain, and served no purpose in treating disease.

Frequently, the patients were the so-called terminal patients. Most lacked sufficient funds to pay for their treatment, and in some experiments a disproportionate number were African Americans.

Reports about the "treatment" of patients by Dr. Eugene Saenger at the University of Cincinnati Medical Center are particularly disturbing. In a project funded by the Pentagon, which commenced at least as early as 1960 and continued for about a decade, Saenger and his colleagues are said to have conducted "what they described as an experimental cancer research project," giving high doses of radiation to some 87 patients. Apparently 25 of these 87 patients died within sixty days of receiving Saenger's radiation treatment.[12]

> It turned out that [Saenger's] entire project was operated under contract to the U.S. Department of Defense, which had contributed funds totaling $650,000. According to the Pentagon's research abstract for the project, it was interested in finding out about potential "reductions in combat effectiveness" in troops exposed to large doses of whole-body radiation of the type that would occur in a nuclear war. These were precisely the kinds of doses that Saenger and his team had been giving to his trusting patients. In its public statements on the matter, the Pentagon maintained that its role in the project was part of its "continuing support of medical

> research." However, the Pentagon sent the results it received from Saenger to dozens of weapons-testers and military officials, but not to a single civilian cancer researcher. The Army medical liaison officer for the project was a veterinarian.[13]

A report issued by a Junior Faculty Associates Committee at the University of Cincinnati concluded that "many patients in this project paid severely for their participation and often without even knowing that they were part of an experiment." Further, "the faculty committee noted that a number of the statements made by the Saenger team in its reports to the Pentagon pointed clearly to the fact that the main reason for increasing the dose over the years was to improve the data—not on cancer treatment—but on radiation injury."[14]

Despite providing "voluminous reports to the Pentagon," in all the years of the project Saenger's team apparently published "absolutely nothing on their whole- or partial-body radiation program as *cancer treatment*."[15]

I will summarize just a few examples of the hundreds of disquieting cases of our misuse of the genie's power.

1. In the spring of 1945 Robert S. Stone, the associate director of health of the Manhattan Project who had an office next to me at Clinton Laboratories, came into my office obviously upset. "Karl, you remember that black truck driver who had multiple fractures in an accident and we rushed him to the [military] hospital?" He went on to tell me:

> Almost all of his bones were broken, and we were surprised he was alive when he got to the hospital; we did not expect him to be alive the next morning so this was an opportunity we've been waiting for. We gave him large doses by injection of plutonium- 239. We were anticipating collecting not just urine and feces but a number of tissues, such as the skeleton, liver, and other organs. This morning when the nurse went into his room, he was gone. We have no idea what happened, where he is, but we've lost valuable data we were expected to get.

Before this conversation with Stone, I had never even heard of the experiment. My work was primarily with physics, not medical or biological studies. I later learned that Stafford Warren (professor of radiology at the University of Rochester, where research was being conducted involving plutonium and human subjects) and Hymer Friedell, a medical physicist, apparently knew about the study. I heard nothing more about this until years later, when I saw in the Knoxville paper a death notice for the black truck driver, whose name I remembered.

2. In 1961 Woodrow Litton worked at the Y-12 plant. During a routine physical two small nodes were discovered under his arm, and cancer was suspected. Litton was sent to the Oak Ridge Institute of Nuclear Studies hospital, commonly known as ORINS. This was reputedly the place to be if you had a cancer problem.[16] During the next four years Litton made five visits to ORINS, where he received radiation treatments, and where he died in 1965. In 1967, while attempting to gather Litton's medical records for his family, his son Gary was told that the records were classified. Not until 1994 were his father's records released, and at that time Gary Litton discovered that Woodrow Litton "was part of one of the most horrifying radiation experiments by the U.S. Government." Litton learned that his father "was used like an animal for experimental purposes."[17]

> My father was given Lanthanum 140, Iodine 123, Iodine 131 (usually given as treatment for thyroid cancer, which my father did not have) and Fe 59. He was given a total of 4500 rads of Co-60 and 150 rads of Cs-137. . . . The most horrifying statement came from a doctor's handwritten notes in my father's medical records that said "Mr. Litton is not responding to radiation Co-60 or total body Cs-137." This was October 28, 1965. Yet on the very next day, the 29th, he was radiated again with 220 rads of Co-60 and at 4:20 a.m. on the 30th of October 1965, he died.[18]

Little did the Litton family know that Woodrow and apparently hundreds of other patients were unknowing participants in a NASA study concerning radiation effects on human beings. Worse yet, "no one doctor at ORINS could agree he even had cancer."

3. During the mid-1950s we learned that some research regarding the effects of uranium exposure was being conducted at Massachusetts General Hospital in Boston. We were anxious to get data, so we contacted the research group and were invited to work with them. I sent my able assistant Ed Struxness to participate. A few days later Struxness walked into my office. I asked, "Why are you back? I thought you'd be in Boston a month or more." He replied, "They are doing things that are very irregular, and I refuse to participate." When I asked him to explain, he told me they were injecting uranium into patients with brain tumors! I put to a halt our cooperation with that program once I learned what was taking place.

4. During the 1940s a secret government project was conducted in which eighteen patients were unknowingly injected with plutonium. It took fifty years for the truth about this deplorable conduct to surface.[19]

5. During 1944 and 1945 the Clinton Laboratories, in cooperation with the Medical Division, placed P-32 beta-emitting plaques on the arms of eighteen employee volunteers to study the skin erythema dose. I became a "guilty" participant in these studies (see chapter 1).

6. In the early 1960s the Oak Ridge Institute for Nuclear Studies fed La-140 to fifty-four hospital patients to determine the uptake and biological half-life.

7. In the mid- to late 1940s Vanderbilt University researchers conducted a study of the gastrointestinal absorption rate of iron by pregnant women, using a tracer dose of an iron radioisotope, iron-59. Four malignancies occurred among the children receiving prenatal radiation exposure, while none was found among the nonexposed children.[20]

8. In the late 1940s and early 1950s the MIT Department of Food Technology conducted studies on mentally retarded teenage boys living at the Fernald State School in Waltham, Massachusetts. Radioactive isotopes were used to trace calcium absorption.[21]

9. In 1962 S. R. Bernard of the ORNL Health Physics Division administered I-131 to himself and other volunteers in his research group. The level of activity neared the maximum permissible concentration for members of the public.

Before 1975 approximately four hundred biomedical experiments of the type described above were conducted. Fragmentary evidence exists of an additional thousand such experiments.[22] One difficulty in uncovering the truth about these incidents is the loss or destruction of important documents by government agencies. A few examples follow.

- In 1973 the director of the Central Intelligence Agency ordered the destruction of files regarding radiation research. The CIA played a role in mid-century committees that planned for human experimentation.
- In the early 1970s the AEC conducted an extensive inquiry into plutonium injection experiments. Although 250 documents were gathered, the DOE (successor to the AEC) has claimed it cannot locate these important documents.
- When Secretary of Defense Wilson issued his Nuremberg Code directive in 1953, he required advance approval of covered human experimentation by the service secretaries. "With limited exceptions, the files containing such approvals have not been located."[23]

Over a twelve-year period from 1951 to 1963 the U.S. military detonated 126 atomic bombs into the atmosphere of the Nevada Test Site. Each pink cloud released high levels of radiation. Tests by the Public Health Service and the AEC, some of them secret, disclosed that crosswinds and downwinds from these tests resulted in skin erythema to many persons downwind and caused the death of hundreds of sheep, cattle, and other animals (fig. 41).

The atomic tests poisoned milk in New England, contaminated wheat in South Dakota and soil in Virginia, and caused damage to fish in the Great Lakes.[24] In 1997 the National Cancer Institute concluded that everyone living in the forty-eight contiguous states between 1951 and 1958 received some fallout from ninety of the nuclear bomb tests in Nevada.[25]

When President Clinton established the Advisory Committee on Human Radiation Experiments in 1994, it was thought that the U.S. government had intentionally released radioactive material into the environment on 13 occasions. As of October 21, 1994, the committee had expanded that number to include 53 radiation warfare experiments and 250 implosion tests involving radiolanthanum.[26]

During some of the Nevada weapons tests veterans were stationed in trenches only a short distance from ground zero. They were instructed to place their hands over their eyes at count five and to wait for the flash at

Fig. 41. Cattle affected by radiation fallout after the Trinity Test. (Photo: James E. Westcott, courtesy of DOE Photography)

count zero. The high gamma dose from the blast enabled the veterans to see the bones in their hands. After the roar of the explosion they were required to leave the trenches and rush toward the blast area. These men are often known today as "the atomic soldiers."

After one tower shot I, along with eight of my fellow physicists, rushed to the blast area in our truck to determine dose rates at predetermined locations in "Japanese" houses we had constructed on the desert floor. Our mission was to determine the doses received by the Japanese survivors. The houses disappeared, but our dose meters, encased in heavy steel balls resting in the debris on the desert sand, were retrieved. During our three-minute time limit, my handheld meter read over 1000 mr/hour at 10 feet from debris of heavy material strewn on the desert floor. I steered clear as far as I could while retrieving the steel balls.

I now shudder to think of the doses our GIs received when they stood near or even sat on this desert debris. Some exposures caused them to die later of "unknown causes." I have offered testimony and made dose calculations for scores of GIs who were suffering radiation injury or who had died from radiation exposure at the Nevada Test Site and during the tests in the South Pacific. Only a few received compensation. The Veterans Administration seems always on the defensive to make sure the victims are not compensated.

Bob Bies was a seventeen-year-old mail clerk in the U.S. Navy who was present during two atomic bomb tests in the South Pacific in 1946. For the rest of his life, Bies suffered cancer. He developed cancer of the stomach and small intestine. He had malignant melanomas on his face. He lost a daughter to cancer. Two of his sons developed cancer. His only surviving daughter also developed cancer. His grandson was born with three kneecaps. All this as a result of mutating genes that he unwittingly passed along.[27]

The saddest case I participated in was that of John D. Smitherman. In 1946 he received large radiation doses at test Able and test Baker, both of which I attended, at Bikini Atoll in the South Pacific. He fought fires on target ships after detonation of the atomic bombs. His health deteriorated rapidly following these exposures. When I last saw him in 1979, infection in both legs had become so severe as to require amputation. His left arm had swollen to the point that it began to burst. The Veterans Administration denied any connection to radiation exposure until 1988, when it awarded his widow benefits. By the time of his death, Smitherman's body was almost consumed by cancers of the lung, bronchial lymph nodes, diaphragm, spleen, pancreas, intestines, stomach, liver, and adrenal glands. In 1989, a year after it had awarded the benefits, the VA revoked them from Smitherman's widow.

When Jimmy Carter became president in 1977, he ordered that the AEC operational records be made public. A team of congressional investigators who studied these records and interviewed participants concluded: "The greatest irony of our atmospheric nuclear testing program is that the only victims of United States nuclear arms since World War II have been our own people."[28]

CHAPTER SEVEN

The Advance and Decline of Health Physics

How many times can a man turn his head and pretend that he just doesn't see? . . . The answer is blowin' in the wind.

BOB DYLAN, "BLOWIN' IN THE WIND"

From my perspective health physics has undergone three stages of development. Although it is impossible to be precise, I have established an approximate time frame for each stage, realizing that history does not always fit into neat chronological packages.

FIRST STAGE, 1943–53

During this early period, the profession realized enormous advances simply because all those concerned shared a common goal—safety. The military-industrial complex had not yet taken its stranglehold on health physics.

My colleagues and I attended many seminars in Chicago in the spring and summer of 1943 dealing with ionizing radiation. Health physics owes a debt to Robert S. Stone, associate director of health of the Manhattan Project. He told us new recruits to the project that we would be working with radiation levels with potential exposures millions of times higher than those previously experienced by human beings. Stone insisted that we implement a conservative approach in determining accepted levels of exposure and accumulated radiation dose. Our decisions on what constituted a safe level of exposure and how best not to exceed acceptable radiation doses developed primarily from "bull sessions" and were based on an innate respect for the unknown.

When the world's first health physicists arrived at Oak Ridge in September 1943, we quickly came to appreciate the danger of ionizing radiation. Stone and Herbert Parker, the first director of health physics at our X-10 facility, left their indelible mark of conservatism on me—a characteristic that became more pronounced with the passage of time. We set what were then considered extremely low levels of permissible exposure.

For the first ten years of health physics from 1943 to 1953, I struggled to raise its status as a science and a worthy profession. I published numerous papers in scientific journals, as well as in popular and religious magazines. I also presented papers at scientific meetings and made a sincere effort to inform the news media whenever security permitted.

One of my key assignments for the entire twenty-nine years I worked at Oak Ridge was to train young employees who aspired to become health physicists. Not only did we train employees for all three Oak Ridge plants, we also educated DuPont personnel responsible for radiation and protection measurements at Hanford. Further, hundreds of military officers from all branches of the service attended our training sessions. Several of these students later held high offices in the military. Others assumed major responsibility for radiation protection at AEC operations. Of these I may mention specifically Jack Healy, who worked at Hanford and Los Alamos, and Morey Patterson, who became the senior health physicist at the Savannah River operations.

No textbooks existed for such training. Therefore, I prepared countless pages of instructional materials and wrote chapters for physics and medical handbooks and other texts.[1] Some of these materials appeared years later in a textbook I coauthored with J. E. Turner entitled *Principles of Radiation Protection* (1967).

In the courses I taught at Clinton Laboratories in 1944, we reviewed thoroughly the methods of producing ionizing radiation and the characteristics of this radiation. I taught my students how ionizing radiation can be attenuated and absorbed by materials of various mass and atomic number and how protection can be provided by distance and limits on the time of exposure.

We studied the scattering of radiation called "sky shine." This term refers to the problem of inadequate shielding in the ceiling of a facility. The gamma and neutron radiation would penetrate into the air above and produce scattered gamma radiation back at ground level.

Our courses also covered "hot cells" where operators used remote-control devices. One to three feet of concrete protected them. It was not practical to use doors on entrances to hot cells where dose rates were millions of roentgens per hour, because the doors, to be effective, would have to weigh many tons. Instead, we used a labyrinth. Physics theory provided most of the information we needed in the design, but we also tried to confirm theory by experiments. We explained to students that the function of the labyrinth was to diminish the intensity of the radiation. This was accomplished by a maze of passageways designed to scatter gamma and neutron radiation many times around corners.

It was not practical to use lead doors both because of the weight and because lead provides very little shielding from neutrons. Concrete serves as a fairly good neutron shield if it contains a lot of water or any hydrogenous material. Since hydrogen (^{1}H) has only one proton in its nucleus and the proton has about the same mass as the neutron, ^{1}H effectively dissipates fast neutron energy when neutrons strike protons like billiard balls.

An important part of our instruction included instrumentation: how to measure alpha, beta, gamma, and neutron (classified as fast, epithermal, and thermal) radiation. It was necessary not only to detect the presence of all these types of radiation, but also to obtain quantitative measurements in terms of roentgens or equivalent roentgens. This required a great deal of skill in developing and constructing new instruments.

It was very important to have instruments capable of measuring only one specific type of radiation in a mixed field of radiation. The risk to human beings from some types of radiation is far greater than for others, depending upon the effectiveness of a shield. Such factors as a shield's density, its atomic mass and atomic number, and its neutron capture cross-section all affect the dose received. For example, paraffin is a good shield for fast neutrons but very poor for X, gamma, beta, and alpha radiation, as well as for thermal neutrons. Lead is a good shield for X and gamma radiation but poor for fast and thermal neutrons. Wood and plastic are good shields for alpha and beta radiation but poor for X, gamma, and neutron radiation. Cadmium and boron are good shields for thermal neutrons but poor for fast neutrons.

Some of these instruments were kept in a fixed location, while others were carried or worn on clothing. I indicated to the students not only the importance of protecting themselves when working with these sources but also the necessity of safeguarding others. For example, a laborer might be safe working at his bench because thick glass protected him from the beta or gamma radiation source in the hood. However, if the back of the hood lacked adequate shielding, the radiation could penetrate into the room behind and seriously expose a secretary sitting at her desk.

We trained students in the basic physics of ionizing radiation and also its biological effects. We did not know about the risks of chronic low-level exposures and therefore could not inform the students about this problem.

During this early stage we were encouraged to disagree, but we also understood our wartime obligation to find the correct answer in haste. Our goals were lofty, and our rather intense "bull sessions" resulted in numerous discoveries.

SECOND STAGE, 1954–66

Women Who Deserve Tribute

Several women scientists made immense contributions to the profession during the second stage of health physics, from 1954 to 1966.[2] For example, ORNL was never the same after the arrival from Los Alamos in the early 1950s of Elda E. Anderson (fig. 42). Anderson relieved me and my colleague Ross Thackery from our overwhelming workload by assuming responsibility as section chief for health physics education and training.

Fig. 42. Elda Anderson, who came to ORNL from Los Alamos in the early 1950s, with me in my office. (Photo: ORNL)

Anderson developed an outstanding training program, along with her very capable associate Myron Fair. Together they developed laboratory and field studies, as well as lectures for the training of health physicists—not only those from the United States, but also many from foreign countries.

In 1955 Anderson and I realized that the training we were providing constituted a graduate-level curriculum. Unfortunately our students received no credit for their studies. To correct this injustice, I worked with Anderson and Herman Roth of the local AEC office. I visited several universities, but only Vanderbilt University showed interest in adding this program to its curriculum. Francis Slack, chair of Vanderbilt's physics department, worked with us to establish the world's first graduate program for the training of professional health physicists.

Anderson devoted herself to the students assigned to the Vanderbilt program. Over the years she assisted hundreds of students. She also visited foreign countries where she offered training programs and added to the prestige of our program.

Later, Anderson, Roth, and I established a second graduate program at the University of Rochester. As time went on, other universities also added graduate programs in health physics. Students from most of these programs came to ORNL to obtain applied and experimental training.

In 1970 Anderson developed acute myelogenous leukemia. After only a few short weeks she died from the disease she so effectively had tried to prevent by teaching health physics.

During her tenure as director of our training program, Anderson often talked to me about her experiences at Los Alamos before joining us at Oak Ridge. It made her shudder to recall how carelessly Los Alamos handled radiation safety, compared to our practices at Oak Ridge. She described with some amusement and chagrin the so-called "tickling of the dragon" at Los Alamos, where they brought a source to criticality very slowly and very carelessly, somewhat similar to the tragic experiment of Lewis Slotin when the screwdriver slipped and cost him his life (see chapter 1). The radiation exposures Anderson received in her time at Los Alamos, in all probability, accounted for her early death.

Three other women—Mary Jane Cook, Isabel Tipton (fig. 43), and Mary Rose Ford—all had an immense impact on the development of health physics. In 1958 I contacted Tipton, my former classmate, who at that time was a physics professor at the University of Tennessee, specializing in X-ray spectroscopy.

I told Tipton that if I knew an element's concentration in food, water, and human urine, feces, and body organs, I could easily develop equations

Fig. 43. Isabel Tipton with Tom Burnett and Doyle Davis (right) in 1958. (Photo: ORNL)

and calculate the maximum permissible concentration (MPC) values for all the radionuclides. I informed her that I had recently spent an entire week in the library, only to find that this information did not yet exist. I asked her if she would be willing to set up a program to develop this information, and she responded with enthusiasm.

Tipton coordinated her study with Cook, who worked in our Health Physics Division. Cook, who had a master's degree in biology, functioned as a section chief, although she did not have that title. She traveled all across the country and abroad visiting pathology laboratories where she collected human tissue samples. Cook would stand right by the pathologist to oversee his work—for example, cautioning him to use glass knives to avoid contaminating the tissue with nickel, chromium, or iron contained in steel knives.

Tipton's study, which was made possible by the work of Cook and others in collecting samples, proved useful in two ways. First and foremost, it provided important information for setting MPC values for many radionuclides. But also this research raised several provocative questions: Why does zinc appear in the human prostate gland or copper in the eye? Why did surprisingly large amounts of cadmium appear in the kidneys of adult residents of St. Louis? Why did the concentration of cadmium in the kidney and liver of thirty-seven subjects from the Pacific Rim exceed, by a factor of 2 or 3, concentrations in forty-nine subjects from the United States, India, and Switzerland?

Tipton was unable to complete all the research she believed necessary, including detailed studies of diets and air contamination among the populations studied. Her work pointed strongly to the need to continue these studies. I was convinced that further research would provide helpful clues regarding the etiology of certain diseases and how to treat and prevent them.

During the twenty years when I served as chair of the internal dose committees of both the ICRP and NCRP (1950–71), Walter S. Snyder, assistant director of the ORNL Health Physics Division, served as secretary of both organizations. Snyder did most of the work of assigning projects to committee members. He also formulated new mathematical procedures to obtain the best values of maximum permissible body burdens of the radionuclides and acceptable MPC values. He searched the scientific literature to obtain the best values of uptake and distribution of radionuclides in the human body, their rate of elimination, and the fractions eliminated via urine, feces, and air, as well as the energy absorbed in each body organ from the various types of emitted radiation as they decayed radioactively in each body organ and while they were eliminated biologically.

All this required a tremendous amount of tedious calculation and library research—work that fell to the lot of Mary Rose Ford, a group leader with a master's degree in biology. In those days we did not have desk computers but only hand calculators. Ford would sit at her desk hours at a time cranking out these data and weeks at a time searching through domestic and foreign publications to find the elusive parameters that were of the highest statistical significance. The internal dose handbooks of both ICRP and NCRP contained thousands of numbers that came from Ford's work.[3]

Formation of the Health Physics Society

After years of effort, health physics began to be recognized as a science and profession. I explored where it would best fit into the scheme of recognized disciplines. I talked with officers of the American Physical Society and the American Industrial Hygiene Association about health physics becoming a branch of these organizations. Soon I lost interest in that approach, because it appeared that if we joined either organization, health physics would lose its separate identity.

I attended the organizational meetings of the Radiation Research Society at Oberlin College around 1953–54, but Alexander Hollander, the principal organizer of radiation biology research programs at ORNL, believed that radiation protection would dilute that group's focus on radiation biology. Finally, after discussions with Anderson and others in 1954, I decided that health physics should establish itself independently. Shortly afterward, lab director Alvin Weinberg urged me to have health physics become a branch of the American Nuclear Society. I rejected his proposal because already many nuclear engineers depreciated the risks of radiation exposure and pressured me to set higher levels of radiation exposure. Thus, over a period of a few months, I firmly committed myself to forming an independent Health Physics Society.

Francis J. Bradley, who did his applied training in our Health Physics Division, proposed that we hold a conference to discuss the official formation of our profession. The program committee elected Anderson as chair. In June 1955 we met at Ohio State University, where Bradley was now on the faculty. There we established the Health Physics Society. I served as the pro tempore president in 1955 and as the first president in 1956–57.

Filled with pride about our fledgling society, I enthusiastically continued to make health physics a profession dedicated to protecting human beings and the environment. I believed that this could be accomplished while still acknowledging and benefiting from the many positive uses of radiation.

By June 1957, when the society met at the University of Pittsburgh, our membership numbered nine hundred. By the following year, when we met at Berkeley, we had established a professional journal, *Health Physics*. Almost from the start it enjoyed a wide international circulation. From 1955 to 1977 I served as editor-in-chief of this journal, which is still recognized as the leading publication on radiation protection. Two years later, at the 1960 meeting in Boston, we formed the American Board of Health Physics and began the process of certification of health physicists. Within five years four hundred health physicists had earned this coveted status.

Formation of the International Radiation Protection Association

Although we intended the Health Physics Society to be international in scope, by 1963 this initial goal had not been accomplished. Scientists from other countries commonly referred to our society as the American Health Physics Society. Large countries, such as the Soviet Union and China, had no members. The founding of two competing European organizations further complicated matters.

As editor-in-chief of *Health Physics*, I tried to stem this tide by forming an advisory board for the journal, composed of health physicists from other countries. This improved our foreign membership only slightly. And soon a third radiation protection organization sprang up in Japan.

In 1963 W. T. Ham, president of the Health Physics Society, asked me to explore the possibility of forming an international health physics organization. My Swedish friend Sigvard Eklund, who at that time was chair of the International Atomic Energy Agency in Vienna, gave me a worldwide list of over a thousand leaders in the field of nuclear energy. I wrote to these individuals to inquire about their interest and was pleased to receive an overwhelmingly favorable response. Scientists in England, France, Italy, Switzerland, Germany, Canada, Japan, the Netherlands, Denmark, and Belgium expressed particularly strong interest. After four organizational meetings, and with the help of J. C. Hart and H. Abbe of our Health Physics Division at ORNL in drafting the constitution, we formed the International Radiation Protection Association (IRPA).

IRPA elected me as its first president in 1966 when the first international congress met in Rome (fig. 44). We voted fifteen societies into IRPA membership at Rome and six more at the next meeting held in Brighton, England.[4] Today IRPA consists of thirty affiliated societies with a membership of over twenty thousand. We sometimes forget that in 1943 there were only five health physicists at the University of Chicago. But did my profession grow in stature as much as in numbers?

THIRD STAGE, 1966–98

Storm Clouds

Even before a nuclear industry existed, H. J. Muller caused repercussions in medical circles as early as 1927, when he suggested genetic damage at low levels of X-ray exposure. The industry did not flinch much when Alice Stewart's studies in the 1950s indicated a cancer risk to children who had received in-utero X-ray exposure.[5] After all, the industry reasoned, Muller's studies were primarily done on flies, and Stewart's studies of pregnant women had limited application to on-the-job exposure in a mostly male workplace.

Several occurrences in the 1970s, most of them unrelated, caused what I call the nuclear-industrial complex to change its view of what the public

Fig. 44. The initial meeting of IRPA in Rome in 1966, where I was elected the group's first president. To my immediate left is P. Caldirola of Italy, the first secretary elected. (Photo: ORNL)

should be told and who should tell it. The nuclear-industrial complex consists of the U.S. DOE and its predecessor agencies, such as the AEC, along with the U.S. military and the nuclear power industry. Two significant events were the publication of a paper by Thomas F. Mancuso, Alice Stewart, and George Kneale on cancer incidence among Hanford radiation workers (1977) and the Karen Silkwood case.

The paper by Mancuso, Stewart, and Kneale, "Radiation Exposures of Hanford Workers Dying from Cancer and Other Causes," demonstrated a statistically significant increase in pancreatic and multiple myeloma cancers as a result of average exposures of only 3 rem received over years of employment at Hanford.

The industry and the DOE were now forced to make a selective analysis by ignoring the equally alarming study of B. Modan and colleagues, "Radiation-Induced Head and Neck Tumors" (published in the *Lancet* in 1974), which showed an increased risk of head and neck tumors at average doses of 9 rem and increases in breast cancer at an average dose of only 1.6 rem. Concerned that its very existence was threatened if the public believed that there was an increased risk of cancer at these low levels of exposure, the nuclear-industrial complex determined that it would vigorously respond to all challengers. With prompting from the DOE, contractors criticized the Hanford study of Mancuso, Stewart, and Kneale.

The celebrated case of *Silkwood v. Kerr-McGee Corporation* in 1979 proved to be the final blow for those who stood to make billions of dollars through the use of nuclear power.[6] This stunning decision caused the general public to question whether nuclear power was indeed "the answer." (For an analysis of *Silkwood*, see chapter 9.)

For the past twenty years, health physics, in its mission to protect and defend persons receiving radiation exposures, has sometimes fallen flat on its face. The following examples (and there are many more) demonstrate why I am convinced that health physics in recent decades has sacrificed its integrity. Certainly there remain some true professionals who will not shade the truth to appease their employers, but they are in the minority.[7]

Hot Particle Problem

An early example of our profession's prostitution occurred with the "hot particle problem" (HPP), which arose during the first five years of operations at the Hanford plutonium-producing facility (1944–49). Small radioactive particles released into the environment caused a substantial health risk to the surrounding population (see chapter 6).

Since the dose from a small radioactively contaminated dust particle varies inversely as the square of the distance from the particle, simple calculation indicates that extremely high local tissue doses of thousands of roentgens will be received by the lung cells close to one of these small particles. Such large doses not only kill most of the cells near the particles but also cause surviving cells farther away to change into primordial cancer cells, the precursors of malignant tumors.

Herb Parker, the senior Hanford health physicist, prepared reports calling attention to this serious problem in 1945 and 1946. For a few years the HPP dominated discussions at division directors' meetings that I attended.

Parker estimates that lung inhalation rates were as high as 2.7 particles per day just outside the T Plant main gate and 3 particles per month as far away as Spokane, Washington, more than 100 miles away. These hot particles contained a mixture of radionuclides, such as Sr-90, Cs-134, Cs-137, Ru-106, and I-131, and undoubtedly some of the particles contained plutonium.[8] Apparently no one conducted Pu-239 measurements at Hanford, but alpha measurements made elsewhere indicated large amounts of Pu-

239, U-238, and U-235 on some of these particles. Two fearless health physicists, Tom Cochran and Arthur Tamplin, broke ranks with their colleagues and published papers in 1974 demonstrating the seriousness of the problem.[9] Few heeded their warning.

The AEC "solved" the HPP when it formed an Advisory Committee of Competent Authority to investigate the matter. Years later I discovered how the AEC selected committee members. A declassified letter written on September 25, 1962, by Paul Tompkins, deputy director of the Division of Radiation Protection Standards of the Federal Radiation Council, to AEC Commissioner Haworth sets forth what I call the Tompkins method of selecting committee members. Pertinent portions of this declassified memorandum follow:

> Memorandum for Commissioner Haworth through Director of Regulation
> Subject: Status Report on Current Activities of the Federal Radiation Council Working Group
>
> 1. It was agreed that current levels of radiation from fallout were too low to impose a practical problem in public health. It was suggested that the Public Health Service come up with its views as to what levels would correspond to enough of a health risk to justify diversion of resources in order to provide protection. If any reasonable agreement on this subject can be reached among the agencies, *the basic approach to the report would be to start with a simple, straightforward statement of conclusions. We would then identify the major questions that could be expected to be asked in connection with these conclusions. It would then be a straightforward matter to select the key scientific consultants whose opinions should be sought in order to substantiate the validity of the conclusions or recommend appropriate modifications.*
>
> 2. Part two would recognize that all of the radiation protection philosophies which have been developed have dealt primarily with problems inherent in the control of sources, where it is recognized that lack of proper care can readily lead to demonstrable injury. . . .
>
> I am proposing to concentrate on the first approach exclusively until this possibility is either dropped, or is far enough along to justify consideration of the second. Any thoughts which you may have, which would be of assistance to me in this project, would be greatly appreciated.
>
> Paul C. Tompkins, Deputy Director
> Division of Radiation Protection Standards

cc: Chairman Seaborg
Commissioner Palfrey
Commissioner Ramey
Commissioner Wilson
Federal Radiation Council Working Group
A. R. Leudecke, General Manager
D. A. Ink, Assistant General Manager
W. B. McCool, Secretary (2)
C. L. Dunham, Director, Biology & Medicine
N. H. Woodruff, Director, Operational Safety

The Advisory Committee proclaimed that the HPP presented no problem after all. In reaching this conclusion, they accepted the meager data they could find that supported what I believe was their foregone conclusion.

The Advisory Committee disregarded early studies of high incidence of in-situ tumors when Sr-89, Sr-90, Y-91, Ce-144, Ra-226, and Pu-239 were injected subcutaneously or intramuscularly into mice, rats, and rabbits, such as the results reported by H. Lisco et al. in 1946.[10] Minute amounts of plutonium-produced cancers at the site of injection and bone tumors occurred frequently in mice, rats, and rabbits injected with plutonium at levels ranging from 0.05 to 5 millionths of a curie per gram of injection—the majority of the plutonium-induced tumors occurring in the spine. One μg of Pu-239 (0.061 μCi) injected locally under the skin would induce fibrosarcomas even though much of the injection dispersed from the site. This frightening and sobering news caused us to increase our efforts to reduce plutonium exposure, but the decision of the advisory committee still stood: the HPP did not exist.

For the cause of justice, the HPP cannot be covered up. Hundreds of thousands of years will pass by before all the evidence can be destroyed. Some of these particles remain in attic heating ducts and furnaces. Some are certainly in the remains of those unfortunates who unknowingly inhaled these radioactively contaminated dust particles. These hot particles will remain in geologic formations for millennia, where future generations of scientists may obtain evidence on the foolhardiness of their ancestors.

NCRP Conflict of Interest

About 1971, the same time the gag order was issued to me in Neuherberg regarding the dangers of the liquid metal fast breeder reactor (LMFBR, see chapter 4), my good friend D. W. Moeller, then president of the Health Physics Society, published his inaugural message, saying, "Let's all put our mouth where our money is."[11] I interpreted his comment to mean that the Health Physics Society would focus on protecting the AEC, DOE, and the nuclear industry from liability and responsibility for their mistakes. He implied that protecting workers or the public was to take second place. At a personal level, I took this to mean that someone like me was an ingrate for expressing opposition to the LMFBR at a time when ORNL was basing its future on building the LMFBR.

For the first time I realized why the AEC and ORNL had given me and health physics such strong support in the past. For nearly three decades I had served as the leading spokesperson for the advancement of health physics. Now I knew the sad truth: protecting employees and members of the public from the harmful effects of exposure to ionizing radiation constituted only a secondary objective of the nuclear-industrial complex. In exchange for the generous economic support given to our profession, we were expected to present favorable testimony in court and congressional hearings. It was assumed that we would depreciate radiation injury. We became obligated to serve as convincing expert witnesses to prevent employees and members of the public who suffered radiation injury from receiving just compensation.

A cursory glance at the National Council on Radiation Protection (NCRP), which set radiation protection standards in the United States, sheds some light on whose hand fed those who set levels of permissible exposure. The NCRP is a U.S. charter organization. I remain one of its thirty honorary members. Most of the 176 NCRP members (as of the early 1990s) also belong to the Health Physics Society.

Past sources of income for the NCRP included the DOE, Defense Nuclear Agency, Nuclear Regulatory Commission, U.S. Navy, American College of Radiology, Electric Power Institute, Institute of Nuclear Power Operations, NASA, and the Radiological Society of North America. In truth, the NCRP relies upon the nuclear-industrial complex for most of its funding other than income from publication sales. Trust me, this fact does not escape NCRP members when they set standards for radiation exposure.

Similar Conflicts within the ICRP

The International Commission on Radiological Protection (ICRP) also has a long history of recommending radiation protection standards. These standards consistently have been incorporated into radiation protection codes of practice, guides, regulations, and laws in most countries. Through the years, major scientists have served on its Main Commission. To mention a few: Sir Ernest Rock Carling, W. Binks, and M. V. Mayneord of the United Kingdom; A. J. Cipriani of Canada; R. M. Sievert of Sweden; and G. Failla and H. J. Muller of the United States.

I also served on the Main Commission; from 1950 to 1971, I was chair of the internal dose committees of both the ICRP and the NCRP. These committees set values for MPCs of all the radionuclides.

Like the NCRP, the ICRP is not free of the grip of the nuclear industry. Since the mid-1970s, most of its members have had as a major objective the preservation of the floundering nuclear power business.[12] A brief examination of some ICRP actions and publications demonstrates why this organization is losing its once high status.

During the early 1960s H. J. Muller, the world's most outstanding geneticist at the time, and I worked hard to develop enough support to convince the ICRP to adopt what became commonly known as the "Ten Day Rule." The purpose of the rule was to prevent women of child-bearing age from damaging their unborn children. Easy to implement, the rule stated that diagnostic pelvic and abdominal X-rays for women of child-bearing age should be delayed except during the ten-day interval following the beginning of menstruation.

The rule was largely based on the widely accepted work of Alice Stewart. In later years the ICRP weakened and essentially rescinded the Ten-Day Rule.

The ICRP also prostituted itself regarding the danger of tritium (^3H), an essential component of the fusion bomb.

Tritium is a very low-energy beta emitter that can be devastating when deposited in human tissue. A helpful, though frightening, analogy to understand the impact of low-energy beta particles on human tissue is to imagine terrorists driving by one's home, spewing out machine-gun fire. If the terrorists' vehicle were traveling at eighty miles per hour, perhaps no more than ten bullets would actually hit one's house. However, if the vehicle were traveling at only five miles per hour, perhaps as many as a thousand bullets would hit the house. In much the same way tritium, with its slow-moving beta emitter, travels through tissue releasing thousands of "bullets"—in this case, knocking electrons out of tissue atoms.

Once we discovered how dangerous ^3H is, W. S. Snyder, my assistant director of health physics at ORNL and also the secretary of ICRP's Internal Dose Committee, joined me in a desperate attempt to increase the "quality factor" of ^3H.[13] Increasing the quality factor results in a proportionate decrease in the MPC. This is significant since the lower the MPC, the more difficult and costly it is for industry and the military to comply.

Stated another way, the higher the quality factor, the lower the MPC of ^{3}H and the safer the working conditions for those employed in facilities handling radiation.

Snyder and I urged that the quality factor for ^{3}H be increased from 1.7 to 4 or 5. We faced strong opposition. Gregg Marley, an ICRP member from the United Kingdom, was at least forthright enough to acknowledge the tight hold of the nuclear-industrial complex over the ICRP. During a meeting of the ICRP main commission, Marley openly admitted that although work conditions would be orders of magnitude safer with the higher quality factor that Snyder and I wanted, such a change would put his government out of the business of manufacturing weapons using ^{3}H. The same would have been true of Los Alamos.

It was particularly disturbing to me to see that the vast majority of Los Alamos radiation workers reaching their hands into gloved boxes were women. (Gloved boxes are a means of protecting a scientist or technician from contaminating his or her hands with radionuclides or from breathing radionuclides. The boxes are about 4 feet long with an 8-inch-diameter opening that has a rubber glove attached to it. Thus the worker can pick up a contaminated object with the glove, inside a hood that is vented to prevent inhalation.) Shortly after I left the ICRP in 1970, it "solved" the ^{3}H problem by lowering the quality factor from 1.7 to 1, where it remains today. It should be no lower than 3 and preferably at 5.

It is unconscionable for any radiation standard-setting body to increase maximum permissible exposure in view of the undisputed evidence that the risk of cancer from radiation exposure is acknowledged today to be fifty times greater than we thought in 1947.

I put on record my displeasure with the ICRP for increasing rather than decreasing permissible exposure levels in the face of this knowledge of increased risk.[14] I firmly believe that these increases in permissible exposure, which were made after I left the ICRP, resulted directly from the profound conflict of interest created by the desire to come to the rescue of the floundering nuclear industry.

The scoreboard shows one fatal cancer per 100,000 person rem expected in 1943, one per 10,000 expected in 1985, and one per 2,000 expected in 1998.[15] To paraphrase folksinger Bob Dylan, how many times can a health physicist turn his head and pretend that he just doesn't see?

Bias of Atomic Safety and Licensing Board

Another example of the disappointing performance of our profession occurred in 1987 and 1988 regarding Three Mile Island waste disposal. I prepared three reports for the National Academy of Sciences and the ICRP indicating that unsafe plans were being made by the NRC, DOE, NASA, and their contractors for the disposal of 2.1 million gallons of tritium liquid waste. I stated that ^{3}H presents a far greater risk than that estimated by the NRC and its contractors—or by me when I was chair of the national and international committees that selected the MPC for the substance. I outlined safer methods of disposal. Along with my three reports I sent copies of other articles as references to support my contentions.

In the winter of 1988 I traveled to Washington as an unpaid witness to testify before the NRC Atomic Safety and Licensing Board. As I approached the witness seat, I looked at the eleven judges seated above me. To my amazement, I was never asked about my three reports. Instead, they only discussed a British article I had included as a reference. As it turned out, the last page of this article, a summary, apparently was not part of the actual publication. However, I had photocopied and submitted this article just as I had received it, including the last page. They kept asking who had sent this to me in this form.

All I could say was, "I don't know. When I open my mail, I throw the envelope away." I had no way of knowing that the last page was not part of the publication.

They treated me like a crook and dismissed me without any discussion of my reports. I left in a daze. How could any group of U.S. government-appointed judges stoop so low as to ridicule a witness for such a trivial matter? If their goal was to have an excuse not to consider a safer way of disposing of more than 2 million gallons of dangerous radioactive material, they had succeeded.

The Galileo Mission

NASA's Galileo mission to Jupiter presents yet another example of the lack of intellectual honesty among health physicists. By way of background, in 1964 a U.S. satellite aborted over the Indian Ocean and sprayed 17,000 Ci of Pu-238 into the earth's atmosphere. Without question thousands of people have either died from this incident or are still carrying Pu-238 particles in their lungs.

Twenty-five years later NASA was planning to place an energy source of 50 pounds of Pu-238 in the Galileo satellite, destined for Jupiter. I sent a letter to the chair of DOE in 1989 pointing out the extreme risk in NASA's plans.

I was concerned that as the satellite circled Venus to gather slingshot energy to speed it to Jupiter, it could easily be exposed to excessive heat because Venus is closer to the sun than the earth, and therefore much hotter (hot enough to melt lead). Excessive heat could damage the navigation system such that on its second pass around the earth it would be thrown off course a few degrees and would burn up in the earth's atmosphere. If this happened, the amount of plutonium in the earth environment might be more than doubled. And the Pu-238 that Galileo was to carry is 152 times more hazardous than Pu-239 on a curie basis.[16]

The DOE's only reply to my letter consisted of a copy of a letter from W. K. Sinclair, chair of NCRP and a member of ICRP, to A. T. Clark of DOE. The letter chided me indirectly for my concern and gave DOE carte blanche for this extremely risky project. In the same year the NCRP received more than half of its budget donations from DOE, the Nuclear Regulatory Commission, and NASA. Fortunately, Galileo stayed on course and reached Jupiter intact. We were extremely fortunate, but I still ask: why do a handful of scientists have the right to secretly risk the lives of thousands of people across the globe?

Massaging the Data

My final example of the dismal state of health physics relates to the vast data base of evidence regarding the survivors of Hiroshima and Nagasaki. Analysis of this population group has enabled scientists to assess more accurately how many cancers will occur in a human population from a given dose of radiation.

The cancer coefficient represents the number of fatal-type cancers that will occur from a unit dose of radiation. During the 1980s corrections to the Radiation Effects Research Foundation (RERF, formerly known as the Atomic Bomb Casualty Commission) data demonstrated the cancer coefficient to be at least three times the value previously in use by the ICRP in setting external exposure limits. Logic dictated that the MPC in turn should be reduced by a factor of at least 3. Instead, most values were increased by the NCRP and ICRP, the standards-setting bodies whose membership was dominated by health physicists.

The industry's concern was dollar-oriented. Once standards-setting bodies recognized a higher dose coefficient, the result would be correspondingly lower MPCs, which would translate into massive funding requests to make nuclear facilities and production plants safer.

Members of the ICRP and NCRP, as well as supporters of the nuclear industry, continue to desperately massage the survivor data to protect their vested interests. Higher cancer coefficients (fatal cancers/rem) are in line with (a) Stewart's analysis of cancer risk from in-utero exposure; (b) the

Hanford occupational exposure data of Mancuso, Stewart, and Kneale; (c) the *tinea capitis* X-ray exposure data of Modan; and (d) other data relating the incidence of cancer from low-dose exposures.

At its September 1987 meeting in Como, Italy, the ICRP acknowledged that the Japanese atom bomb survivor data indicated that the cancer coefficient was at least three times higher than it had previously conceded. The ICRP went on to state, "This information alone is not sufficient to warrant an immediate change in the dose limits."[17]

A RAY OF HOPE

I am still very grateful to the handful of health physicists who joined me in the early period before 1950 in the fight to reduce excessive exposures from medical and dental X-rays. They included H. J. Muller, H. Blatz, H. Jones, E. R. Williams, J. S. Laughlin, F. M. Medwedeff, and L. H. Heanpleman.

We deplored the equipment and techniques commonly used in medical diagnosis and especially in the mass programs for chest X-rays. Our measurements indicated that children—routinely X-rayed by the thousands—were receiving an average surface dose of 2000–3000 mrem. At the same time our facility at ORNL under the direction of Tom Lincoln delivered an average of only 6 mrem per chest X-ray. There was no justifiable reason for children to receive a dose 500 times greater except to save a few pennies per X-ray.

After nearly twenty years of frustrating failures, a highlight of my life's work occurred in 1968 when President Johnson invited me to the White House to witness his signing of the bill to reduce such excessive medical exposures. As I watched him sign Public Law 90-602, "Radiation Control for Health and Safety Act of 1968," I realized that thousands of lives would be saved. But why did it require an act of Congress for the medical profession to institute responsible chest X-ray programs?

In the past decade a handful of health physicists, physicians, and others have been courageous enough to step forward and challenge the standards-setting bodies and the nuclear establishment in court, in defense of justice to those injured from excessive radiation. This is a faint ray of hope. A few of these heroes whose names readily come to mind are Alice Stewart, John Gofman, Ted Radford, Tom Cochran, Dean Abrahamson, Ralph Nader, Clair Nader, Jon Cobb, Victor Fenton, Jack Geiger, Carl Johnson, B. Modan, Tom Mancuso, George Kneale, Karen Silkwood, Joe Muller, Walter Snyder, and Art Tamplin.

CHAPTER EIGHT

Ecology and Nuclear Waste Disposal Studies

Look deep, deep into nature, and then you will understand everything better.

ALBERT EINSTEIN

As the director of the Health Physics Division of ORNL, I soon realized that protecting humankind and the environment from the hazards of radiation required more than a knowledge of physics. I needed answers to questions involving chemistry, biology, ecology, geology, sociology, medicine, engineering, etiology, hydrology, soil chemistry, mathematics, meteorology, seismology, ichthyology, and other fields.

My two assistant directors, Walter Snyder and Ed Struxness, kept me afloat in this vast sea of unknowns. Three particularly gifted scientists on my staff—Rufus Ritchie, Sam Hurst, and Bob Birkhoff—were also invaluable. Further, my section chief Jim Turner possessed a remarkable combination of theoretical and experimental knowledge of physics, which proved valuable time and again. Turner also served as a coauthor with me on, among other things, our book *Principles of Radiation Protection* (1967, 1973).

Every time I grasped for an answer it dissolved into yet another question. In response I employed young scientists of many diverse professional backgrounds to conduct theoretical and experimental research. They provided answers to a wide variety of problems.

As a fail-safe measure, I established advisory committees composed of some of the world's leading scientists to lend guidance to these bright but inexperienced professionals. These advisers included John Wheeler (nuclear physics), Eugene Odum (ecology), and Able Wolman (sanitary engineering). All research projects focused on two central objectives: understanding the interaction of radiation with matter, and learning how human beings and the environment can be protected.

We engaged in more than twenty different categories of research. Beginning with a handful of scientists in the fall of 1943, the Health Division numbered over two hundred when I left ORNL.

Annually, the ORNL Health Physics Division prepared "progress reports." These reports contained a wealth of experience about such matters as the problems of handling radioactive materials, radiation ecology, radioactive waste disposal, radiation monitoring, interaction of radiation with matter, and permissible exposure levels. They summarized the results of millions of dollars of research funded by the American taxpayer.

Fortunately, I retain all but two of these progress reports in my personal library. Obtaining the two missing reports proved to be a major challenge. All these reports were placed under lock and key in a special section of the ORNL library.[1] Much of the material for this chapter derives from these restricted reports.

ECOLOGICAL STUDIES

AEC Funding Difficulty

The phrase "ingenuity fueled by desperation" adequately describes the dilemma I was placed in as a result of the AEC's early reluctance to adequately fund ecology and waste studies. We turned to other resources and sought the assistance of key experts, such as geologist P. Stockdale of the University of Tennessee and G. DeBuchananne of the U.S. Geological Survey. Scientists from the Army Corps of Engineers and the Public Health Service also shared their expertise. This approach allowed us to struggle along for ten years. We eventually won over our local management and the Division of Biology and Medicine of the AEC to the realization that our studies on radioactive waste and radiation ecology were of paramount importance. By the time I retired in 1972, we had obtained strong support. Considerable credit for indoctrinating our funding source belongs to a local AEC supervisor, Sam Shoup.

Studying the Food Web

Louis Krumholtz first performed our radiation ecology studies. They were later augmented by the work of Orlando Park (fig. 45), Struxness, and Stan

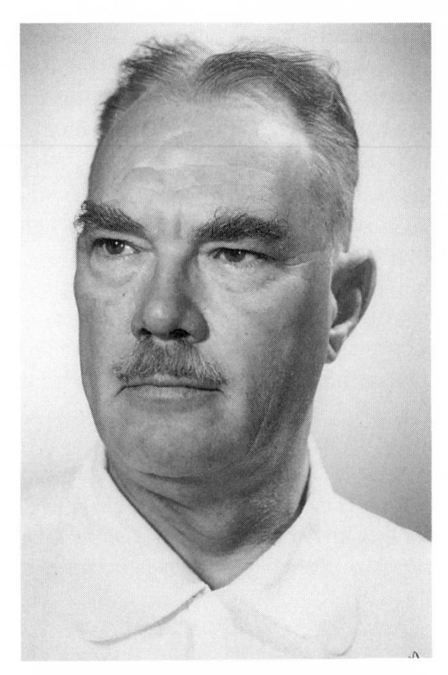

Fig. 45. Orlando Park, who became one of the founders of radiation ecology as a result of his work with the ORNL Health Physics Division. (Photo: ORNL)

Fig. 46. Stan Auerbach, a former student of Orlando Park at Northwestern University, who in 1954 was chosen to head our radiation ecology section. (Photo courtesy of Stan Auerbach)

Auerbach (fig. 46). Although this is not commonly recognized, Struxness and Park became the founders of radiation ecology as a result of their work with the ORNL Health Physics Division. A former student of Park at Northwestern University, Auerbach was chosen to head our radiation ecology section of the Health Physics Division in 1954. Park proved instrumental in encouraging other AEC facilities to conduct studies similar to those under way at ORNL. He promoted widespread interest in this new field among members of the Ecological Society of America. As a consequence, in 1955 the AEC Division of Biology and Medicine established an ecology program under the direction of J. N. Wolfe. By 1956 two promising young scientists, Dan Nelson and Jerry Olson, had joined our radiation ecology section, which continued to prosper until it became a separate division, the Environmental Sciences Division, about the time of my retirement.

During 1955–56 Auerbach extended a lifelong study of tree holes by Orlando Park. When a tree limb breaks off, often a hole is formed in the trunk. Over time it becomes filled with punk. Sometimes woodpeckers help in forming these tree holes. Extremely interesting to the ecologist, tree holes provide a virtual universe of many kinds of living organisms in a state of balance or transient equilibrium.

Iodine-131 was one of the numerous radionuclides we followed in our research. After leaving ORNL (via stacks or White Oak Lake), I-131 was deposited on grass, which was eaten by cows, which produce milk, which is given to babies and young children. Unfortunately, the I-131 found its way to the thyroids of these youngsters. Our waste ponds served a useful purpose, since they held up much of the iodine radionuclides long enough to make detection of their presence in the environment difficult.

The leaves of trees in portions of the Oak Ridge Reservation contained radioiodine from what was taken up by the roots of the tree and from what was occluded on leaf surfaces. In the fall the leaves dropped to the ground and were acted upon by the bacteria, fungi, arthropods, and all sorts of insects, ending up in field mice, rabbits, deer, and human beings. Each radionuclide in these cycles presented new puzzles that were exciting to solve. Some made us a bit apprehensive and more cautious in setting the levels of MPC of radionuclides in air, water, and food.

Louis Krumholtz discovered that fish in White Oak Lake (as well as the many life forms they depended upon for food) took up selectively certain radionuclides. A particularly alarming finding was that the body organs of some species of fish concentrated Sr-90 by a factor of 10,000! Much to Krumholtz's and my disgust, government policy prohibited the open publication of these studies. All of his research findings were conveniently filed away (hidden) to "await completion of the study." It is to be hoped that the federal government will locate and make available to the public Krumholtz's two-volume report, submitted in the mid-1940s.

Because of its flimsy clay structure, White Oak Lake Dam nearly washed out on several occasions during the first twelve years of operation. On more than one occasion I spent hours at the dam when water began to run over the top. With bulldozers and shovels we filled in enough dirt to avoid a disaster. Our geology consultants warned us that we could expect an occasional downpour that would wash away this makeshift dam.

By October 1955 our only feasible option was to slowly drain the lake. This carefully selected time period coincided with increased flow from rain on the Clinch River. Thus, sufficient dilution was achieved so that we did not exceed the MPC values I had set for exposure to the public. Those values amounted to one-tenth the MPC for our workers. Needless to say, I would not make such a release of radionuclides into the public environment today.

After White Oak Lake was drained, Struxness and Wolfe concluded that the contaminated lake bed presented a near perfect laboratory setting for an ecological study that would interface more closely with human beings and the environment than tree holes. As a result, in 1956 an entomologist, A. Crossley, and J. B. Lackey, an aquatic ecologist and planetologist, joined the ecology section to study the rapidly vegetating lake bed.

After analyzing 250 core samples taken from the bed of former White Oak Lake, it was determined that as of December 1962, the soil contained 1037 ± 88 curies of ruthenium-106, about 704 ± 34 curies of cesium-137, 152 ± 9 curies of cobalt-60, and 15 ± 1 curies of strontium-90.

The Oak Ridge Reservation was a 30,000-acre forest of second-growth trees, including pines, dogwood, sweet gum, poplar, oak, maple, and sycamore. Jerry Olson provided valuable contributions in his area of specialization, forest ecology.

Eugene Odum, a leading U.S. ecologist and a principal consultant of our radiation ecology program, convinced Struxness and me to expand our ecology program beyond the reservation. Our expanded program included the Clinch River below the outfall of White Oak Lake and Watts Bar Lake, which at top level backed up the Clinch slightly above White Oak Lake outfall. Dan Nelson, one of Odum's doctoral students at the University of Georgia, led our radiation aquatic program. Eventually he became the associate director of our radiation ecology program and, along with Auerbach as the director, played a major role in our entire ecology program.

After Nelson joined our program and the Clinch–Watts Bar studies were launched, the health physics radiation ecology program expanded rapidly as we recognized the need for an all-encompassing program. It was then (1959–61) that Steve V. Kaye, a mammologist, Martin Witkemp, a soil microbiologist, and John Witherspoon, a plant ecologist, were added to our team.

A Lesson Learned from H. J. Muller

The worm known as *Chironomus tentan* unlocked some mysteries about how radionuclides are recycled. This worm lives on the bottom of rivers and lakes. We gathered samples from the bottom of White Oak Lake and the Clinch River, seeking answers to these questions: (1) Has this part of the ecosystem sustained damage from radionuclides released into the environment by ORNL? (2) Does *C. tentan* clean the mud of radionuclides by returning them from the bottom mud to the water and thereby increase human exposure? (3) Do clues exist relative to the radiation-induced genetic mutations in human beings after thousands of years of exposure to ionizing radiation?

My dear friend Joe Muller (fig. 47) caused me to be particularly interested in the last question. During meetings of the ICRP from 1962 to 1966, genetic damage received primary focus in setting levels of maximum permissible exposure for radionuclides. This was not surprising since Muller was an ICRP member. In order to spark debate, I would sometimes take a contrarian view to "stretch" Muller's thinking. Muller simply would not budge on his position that the limits of MPC should be set first to avoid genetic damage and second to prevent somatic damage.

The greatest concern Muller expressed to me concerned nonvisible mutations. His studies on flies showed that for every visible mutation, 10,000 nonvisible radiation-induced mutations occurred. Muller brought his point home to me by emphasizing that in human beings this means the inability of a species to reach full potential—fewer great scientists, philosophers, artists, musicians, or statesmen.

We examined cells from the salivary gland of *C. tentan* samples taken from the bottom of White Oak Lake and the Clinch River. We compared these cells with *C. tentan* cells taken from the bottom of Ten Mile Creek, a nearby creek uncontaminated by radionuclides. Had we not been so adamant in our conviction that radiation does more damage than most scientists realize, we would have missed a startling observation.

Fig. 47. My dear friend H. J. "Joe" Muller, Nobel Prize–winning geneticist, who taught me about the nonvisible mutation caused by radiation. (Photo courtesy of Indiana University Archives)

Three endemic (native) inversions were studied as an index of polymorphism, a characteristic believed to enhance chances of survival and general fitness, but no differences were found in the two populations. It appeared that radiation exposure of about 230 rem per year did not impair the fitness of these worms. Thanks to mentors like Muller, however, we refused to jump to broad conclusions.

Four chromosomes are normally found in the nucleus of the salivary gland cells of *C. tentan*. However, an unusually large percent of specimens from White Oak Lake and the Clinch River possessed only three chromosomes in the cell nuclei. Unfortunately, by the time I left ORNL, the consequences of these chromosomal changes had not been determined. I expected that even the most avid nuclear supporter from the scientific community of geneticists would not want to see a similar insult to human cells.

We certainly did meet one of our objectives, which was to determine whether clues exist in nature relative to radiation-induced mutations in human beings. We also answered our first question by proving that this part of the ecosystem (river bottom) had sustained damage from radionuclides released into it by ORNL. A dose of only 17 rem caused three times the variety of aberrations one would expect and also produced ten new types of aberrations for this organism.

Our studies and those of other investigators led to an affirmative answer to the question of whether *C. tentan* worms cleanse the mud of radionuclides by returning them from the bottom to the water, thereby increasing human exposure. *C. tentan* and other tubificid worms represented the most abundant bottom organisms in White Oak Lake and the Clinch River. In feeding, they bury their heads deep in the bottom mud with their tails waving in the water above. Thus they bring into the water or recycle radioactive bottom sediments as they defecate into the water. The pellets of feces were found to be filled with digested radioactive waste containing such radionuclides as Pu-239 (which has a half-life of 24,110 years). Rather than remaining buried in the mud, *C. tentan*, simply by doing what nature intends it to do, causes radionuclides to be constantly returned to the food web. Thus we answered another question: *C. tentan* does cleanse the mud of radionuclides and in so doing causes a substantial increase in human exposure.

To me, a nongeneticist, our most interesting finding was the increase in the variety of mutations but not in the number. It almost gave me the jitters when I realized that in our five-year study these worms had passed through 120 generations. For human beings, 120 generations would take us back to 2000 B.C. These worms each received an average dose of 17 rem, and this study showed some of the genetic consequences resulting from exposure. So yes, clues do exist regarding radiation-induced mutations in human beings thousands of years after exposure.

Clam Shells as Historians

Our radiation ecology program also discovered the value of clam shells as historians. Clam shells serve as recordkeepers of past discharges of radioactive materials and other wastes into the water system. Each season the clams lay down a new shell, which contains a sampling of all the minerals during the past time period. These shells consist primarily of calcium carbonate, $CaCO_3$.

Clams are relatively immobile, so each clam shell is essentially a recorder of minerals in the water at a specific location. A collection of clams from many sites provides a geographical history of an entire area. Of course, radionuclides like calcium-45 (162.7 d), carbon-14 (5,715 y), strontium-90 (29.1 y), radium-226 (1,599 y), and barium-140 (12.75 d) show up in larger concentration in the shell layers, but other radionuclides, such as iodine-129 (1.7×10^7 y) can be detected and serve as tracers for iodine-131 (8.04 d) in an indelible record for millions of years.

Thousands of curies of radioiodine in the Three Mile Island reactor at the time of the accident there have never been accounted for. Clam shells in the nearby Susquehanna River may hold the history of what happened to that radioiodine. This biological instrumentation is by no means limited to a record of radionuclide releases into the water. It is applicable also to nonradioactive chemical releases.

In these studies we sliced the clam shells at right angles to the shell layers. Each layer was studied microscopically, much as one might examine tree rings. One can obtain a gross picture of the radioactive releases into the river by placing the slice of the clam shell on a photographic film for an appropriate time in a darkroom. Studies similar to those of clam shells were also conducted on fish scales.

Our radiation ecology studies resulted in observations that raised new questions we were never given the time and resources to answer. For example, why does the wasp select radioactive mud to build its little mud houses? They are so good in their selection process that we found no difficulty locating their nests using a Geiger counter.

Ultimately nature itself served as a reliable witness of the type and extent of radionuclides that human beings were carelessly releasing into the public domain.

Fast Neutron Damage

A study of the impact of fast neutrons on pine trees soon made me realize how dangerous this type of radiation can be. Three-year-old pine trees (*Pinus* Virginia Mill) were given a wide variety of radiations. These sturdy trees did not reach a mortality rate of 100 percent until they had received an acute dose of 1200 rads of gamma radiation. Yet the same type of trees could only withstand a 100 rad dose of fast neutron radiation before reaching 100 percent mortality![2]

DISPOSAL STUDIES

Studies Using Bedded Salt Formations

In 1958 and 1959 we began studies on new methods for disposal of medium- and high-level radioactive wastes. From 1959 through 1972 we carried out extensive theoretical and experimental research on disposal in bedded salt (NaCl) formations.

Radioactive waste disposal in natural salt formations may present problems, depending upon the extent to which the structured properties of salt are affected by chemical interaction, pressure, temperature, and, of course, radiation. Special precautions must be taken to use the best method of handling any gases that may emanate.

Our early studies indicated minor effects of radiation on the structural properties of rock salt. Our calculations and experience indicated that although Wigner energy is temporarily stored in the salt, the rising temperature caused by the radiation itself is sufficient to slowly release this energy and prevent an explosion or sudden rise in temperature. Our theoretical and experimental data told us that the temperature of the radioactive wastes stored in salt cavities could be kept within acceptable limits by controlling the age of the stored wastes as well as the spacing and size of the cavities.

We conducted numerous studies for over a decade both in our health physics laboratory at ORNL and in situ in the Carey Salt Mine in Kansas. Large rooms dug out in an abandoned salt mine served as our in-situ laboratories.

During the nineteen-month mini-operation the average dose to the salt under study was about 800 million rads, with a peak of 9 billion rads. Doses dropped off rapidly with distance and radiation attenuation in the salt. As expected from our ORNL laboratory studies, the salt suffered no significant chemical or structural damage due to the radiation.

We detected small amounts of organic peroxide when the salt reached a temperature of 175°C. This should be of no consequence in an actual disposal operation. Ultimate doses to the salt with actual disposal operations would exceed 10 billion rad, but the mass of salt damaged at a few inches out from the waste units would be so small that this would not be expected to produce any adverse consequences.

We concluded that the effects of stress, temperature, radiation, and other variables would not require the use of conventional roof-bolting techniques unless a shale layer was too close to the roof. Thus, the floor of an entire room in the salt mine would be available for wells to hold the canisters of high-level waste or spent fuel elements. After the entire floor of each room was filled and acclimated to these "hot" canisters, our plan called for it to be backfilled with crushed salt. Within a few decades all voids would be closed due to plastic flow of the salt, and the waste would be tightly encased in a solid mass of salt.

If the waste containers were made of mild steel, then only generalized rusting would be expected, and the integrity of the waste containers should be maintained for an indefinite period of years. Even if the containers perforated, no problem would be expected, since there should be negligible gas pressure inside the container. If some gas emanated from the damaged container after thirty years or so, the rapidly annealing crushed salt above would act as a good filter.

We estimated a surprisingly low cost for such a high-level waste facility. For example, a salt mine facility handling the entire high-level waste output from the U.S. nuclear power industry in the year 1972 would have required a capital outlay of about $10 million. The annual cost would have been about $2.5 million for reactor wastes that had cooled in the reactor water pool for only one year and about $1 million for the waste that had cooled for thirty years.

I considered our studies on disposal of reactor wastes in natural bedded salt only preliminary. Many areas required further research. Disposal in other rock formations might prove to be a better choice, but there seemed to be several basic advantages of salt. For example, water is the principal substance that could cause migration of waste over periods of millions of years. These salt formations would not exist if they had not been isolated from water for many millions of years. Also, salt, although an essential substance for human existence, is cheap and plentiful. Salt would never be a rare substance of such value as to tempt future generations to dig for it and inadvertently encounter a radioactive waste storage area.

Hydro Fracture Studies

Hydraulic fracturing was another promising method of high-level radioactive waste disposal that we developed. We used this method, along with our pit and pond disposal system, for several years as the major means of disposing of our low- and intermediate-level wastes.

Hydraulic fracturing involves the selection of an area underlain at a depth of at least several hundred feet with a geological formation, such as shale, that can be fractured in a horizontal plane. Fortunately, Conasauga shale lies under the ORNL area. Several experiments indicated that this local formation met our basic requirements and was not near the water table level.

Having located a suitable shale formation, we dug holes to a depth of several hundred feet. We lowered into each hole a high-strength steel casing with a welded plug at the lower end and cemented the casing into the holes. On the ground surface, large mixers and blenders mixed portland cement with the radioactive waste. We lowered a blasting device to the desired depth, where a slot was cut in the steel pipe. Next, we pumped water under high pressure into the hole to crack the formation. Then, again under high pressure, we pumped the mixture of cement and radioactive waste into the well. When we completed this injection, we sealed the slot in the casing and cut a new slot in the pipe casing a few feet above. When another waste batch became available, we repeated the procedure.

We injected a total of 1,444,826 gallons of waste mixture into the Conasauga shale at our test site. Altogether in these experiments we disposed of about 1.7 million curies of radioactive waste, of which over 200,000 curies were Cs-137. In the course of these studies on hydraulic fracturing, we developed many additives, which we mixed with the waste-cement mix. These improved the technique by preventing flash settling and gave the desired viscosity, mixture uniformity, rate of settling, and strength of hardened mixture. With the proper combination of additives, the mixtures of concrete and waste resulted in the grout settling in large pancakes, hundreds of meters in radius. These underground pancakes were each about one-quarter of an inch thick. We considered this the ideal arrangement because it provided sufficient area to dissipate the heat from radioactive decay, thus preventing melting of the shale or the formation of steam.

Deep Well Studies

We conducted a detailed study of a waste disposal system often used in the oil industry—namely, deep well injection, in which the waste is forced into formations thousands of feet below the surface. This method seemed ideal for the disposal of large volumes of tritiated water (water containing ^{3}H), such as those later accumulated in the cleanup following the Three Mile Island reactor accident. But it was not suitable for our wastes because, in such operations, all the wastes have to be dissolved completely in the water. Otherwise the deep well would soon plug up and stop the flow.

Other Disposal Studies

One of the most intriguing, but cost-prohibitive, disposal methods that we briefly investigated involved transporting highly concentrated radioactive wastes by satellite to the moon or the sun. The malfunctioning U.S. rocket that reentered the earth's atmosphere and incinerated its 17,000 Ci of Pu-238 into the upper atmosphere over the Indian Ocean in April 1964 impressed upon us that the satellite method should not be further explored. Our calculations indicated that millions of persons might inhale these particles in their lungs. A dose close to a submicron particle would be hundreds of rem per day.

Disposal in caves in a granite mountain also looked attractive. We decided that this waste should be in a solid or solidlike form, such as glass, and in a configuration that would dissipate the heat without changing the form or structure of the waste or its encapsulating material. Ideally it should be encased in something like gold. However, we concluded that this or any other very valuable metal should never be used because of the temptation of thievery. Thick stainless steel might be the best choice. If possible, the structure should be very heavy and an object extremely difficult for anyone to remove. By the time I retired, we had never commenced experimental studies of disposal in caves.

Since my retirement in 1972 the Health Physics Division has been divided into many parts and parceled out to other laboratory divisions where radiation protection is not the primary objective. Is limiting health physics research cost-effective? The answer can be found by a visit to the former Soviet Union, where the consequences of ignoring health physics concerns are only too visible today.

CHAPTER NINE

The Genie Goes to Court

All the world's a stage,
And all the men and women merely players;
They have their exits and their entrances;
And one man in his time plays many parts.

WILLIAM SHAKESPEARE, *AS YOU LIKE IT*

Silkwood v. Kerr-McGee Corporation (1979) and *Allen v. United States* (1984) constitute the two most significant radiation cases of the century. Dr. John Gofman and I served as the primary expert witnesses in both of these cases (figs. 48, 49).[1]

The controversy with Kerr-McGee concerned a young woman, Karen Silkwood, who worked at a Kerr-McGee plant in Cimarron, Oklahoma, that manufactured plutonium fuel rods. Karen, a lab technician, raged and hollered about the plant's lax safety conditions. She complained about routine spills, a cavalier attitude on the part of management, and the fact that no one informed the workers that plutonium causes "cancer."

Upset that she had become contaminated, Silkwood embarked on a campaign to expose Kerr-McGee. An elected member of a union bargaining committee, she persuaded the union to file before the AEC thirty-nine charges of lax safety conditions.[2]

Silkwood insisted that Kerr-McGee had doctored photomicrographs of fuel rods in order to pass inspections. If Silkwood could prove her suspicions, the information would be leaked to *New York Times* reporter David Burnham. Silkwood labored into the evenings examining photomicrographs of fuel rod welds for markings of black felt-tip pen, used to cover

Fig. 48. John Gofman, a codiscoverer of uranium-233 along with Seaborg and the first person in the world to isolate plutonium. Gofman and I served as key expert witnesses in the Silkwood and Allen cases. (Photo courtesy of John Gofman)

up defects. She informed a friend that 40 pounds of plutonium was missing from the facility.

On November 5, 1974, Silkwood discovered contamination on her right wrist during a routine monitoring. Company personnel found plutonium traces in her nose. Two days later fecal and urine samples revealed plutonium inside her body.[3] That afternoon she learned that plutonium had been found in her apartment.

Convinced that someone was deliberately contaminating her, Silkwood showed a friend a reddish brown notebook that she said contained proof that Kerr-McGee had falsified its records. She intended to hand over the evidence to the *Times* reporter that evening. Shortly after that discussion, Karen Silkwood died mysteriously.

The Oklahoma Highway Patrol concluded that she fell asleep at the wheel of her Honda. Many people disagreed, pointing to contrary evidence. Tire tracks of her Honda led partly down an embankment and then sharply back to the road into a bridge abutment. How could she have steered her car back to the road had she been asleep? Also, why were there fresh scars and paint marks on the back bumper of her Honda, and why was there other suspicious evidence that was revealed during the congressional hearings where I testified? The evidence that Silkwood had shown her friend disappeared, despite the fact that a police officer remembered seeing papers at the scene, picking them up, and putting them in the front

Fig. 49. Karl Z. Morgan in 1979, the year of the Silkwood case. (Photo courtesy of ORNL)

Fig. 50. Gerry Spence, trial counsel for the Silkwood family in Silkwood v. Kerr-McGee. *(Photo © 1995 by D. J. Bassett)*

seat. The file folder and the reddish brown notebook never reappeared. The congressional hearings caused me to conclude that Silkwood had been murdered. The Silkwood family vigorously pursued a private civil action against Karen's employer.

The *Silkwood* case was set for trial in March 1979 and brought together diverse personalities. Gerry Spence, the "Cowboy from Wyoming," fortified his national reputation during the course of the trial in his capacity as the attorney for the Silkwood family (fig. 50).[4] Behind the cowboy image stood a talented trial lawyer. Spence not only believed in this case, he understood how to effectively communicate to a jury.

Bill Paul, an experienced corporate trial lawyer from Oklahoma, represented Kerr-McGee. Paul's demeanor was just the opposite of Spence's. Paul always looked as if he had just walked out of a high-dollar traditional men's clothing store that sold nothing but gray suits, while Spence would wear his cowboy hat to court.

Judge Frank Theis, chief judge for the District of Kansas, was specially assigned to hear the case in Oklahoma (fig. 51). A tall man with a booming voice, at times he was as witty as Will Rogers. After observing him in the

Fig. 51. Judge Frank G. Theis, who tried the Silkwood *case. (Photo courtesy of Van Dusen Photograph)*

courtroom for thirty minutes, I felt confident that the Silkwood trial would be a fair one to both sides.

Spence delivered his opening statement on March 7 before a packed courtroom. Presented before the jury hears any testimony, the opening statement gives the jury an overview of the case. Spence raced to establish his credibility with the jury before Paul had the opportunity to present Kerr-McGee's version of the facts. "Let's be up front about this—this is a case about money. That sounds crass and gross, but that's what it's about. It's true, the case is also about pain and death and fear and terror and panic, and it may be a case about the future survival of the American people, and it's a case about damages, whether or not there will be damages of sufficient size to make Kerr-McGee and other companies perform their duties in a way that will allow the American people to survive."[5]

The first witness Spence called was John Gofman, a scientist who held degrees in both chemistry and medicine. Along with Glenn Seaborg, Gofman codiscovered uranium-233, and he also was the first one to isolate plutonium. In spite of these achievements, Gofman has yet to receive the recognition due him; in my opinion, he is one of the leading scientists of the twentieth century.

When Spence asked Gofman whether he was "acquainted with plutonium," Gofman replied that in 1940 as a graduate student in nuclear chemistry he joined Seaborg and Arthur Wahl as "the world's third chemist to work on plutonium" (689).[6]

Spence then read to the jury portions of Gofman's impressive curriculum vitae. As part of his background, Gofman stated that in 1974, at the personal request of Ernest Lawrence, the founder of the Lawrence Livermore Laboratory, he agreed to serve as medical chief of the Livermore Lab. Later he became the first head of Livermore's Bio-Medical Division. This first day of testimony made it clear that Gofman possessed credentials specifically suited to the *Silkwood* case.

The following day, March 8, Gofman resumed the stand. Walking over to the courtroom chalkboard, he explained the nature of uranium, alpha particles, neutrons, atomic numbers, and plutonium. Plutonium, he stated, constantly emits "ripping bullets" (called alpha particles) that travel thousands of miles per second (727–28).

To make the jurors appreciate the dangerous nature of plutonium, Gofman informed them of what occurs when such a minute amount as a nanocurie is inhaled into the lung:

> Two thousand times a minute these bullets, alpha particles, are coming out . . . delivering 5 million of those volts of energy, each one. So, it's a fantastic projectile. The alpha particles in the lungs, it is hitting right through the cells of the lung with 2.5 million times the energy that you would get from a carbon burning. So, you see, expecting that your cells are not going to be damaged by that would be about the same expectation when somebody might talk to you and say: "Well, a small amount of this won't hurt you." That is such an absurd nonsense notion that one wonders how anybody could think of it. (731–32)

Gofman used the analogy of "taking a beautiful color television set, surrounding it with machine guns, ripping bullets into it for an hour, and then saying, "I expect that television set will function just fine" (733). He testified that every gram of human tissue contained between 100 million and 1 billion cells. He told the jury that all cells coordinate with each other in order to function properly. Every tiny cell contains a library of instructions that would fill many volumes of books.

The jury hung on Gofman's every word as he explained how alpha particles rip into a cell and create utter havoc with the library. Gofman stated that once the instruction book in the cell library is damaged, altogether new cells, cancer cells, are formed. "Cancer cells go off on their own. They live by their own law, because the instruction book has been damaged" (738).

"The amount of plutonium, from my research and [I'm] in agreement with others, what it takes, to guarantee that a human will go on and show a clinical fatal lung cancer, is very small" (741). Gofman explained that only 7.3 micrograms of plutonium deposited in the lung will "guarantee a fatal lung cancer" (742).

Gofman testified that he had studied the autopsy documents of Karen Silkwood and confirmed that her lungs contained 1.3 times the amount of plutonium "required to guarantee her lung cancer" (746).

Spence didn't stop with fatal lung cancer. He wanted the jurors to know that alpha particles cause other kinds of damage. Gofman answered that once damage occurs to the intricate "instruction book" in the nucleus of a cell, all kinds of adverse health consequences can result. He explained that fibrosis of the lung is caused by scarring due to the intense radiation and that it makes breathing nearly impossible, causing death from asphyxia (751).

Gofman also noted that plutonium enters the bloodstream and occasionally is deposited in the female ovary or the male testis. If either the testis or the ovary is bombarded with plutonium alphas, a defective sperm or ovum results. Gofman made it clear that the fertilized ovum sometimes has enough faulty instructions to make a living being, born with one or more of hundreds of types of defects (753). The jury was told that these various defects can lead to premature death, mental retardation, hemophilia, cystic fibrosis, and an increase in susceptibility to heart disease, high blood pressure, and arthritis.

Spence asked Gofman to comment on Paul's assurance in his opening statement that plutonium exposure inside the Kerr-McGee facility was within safety limits set by the AEC. Gofman replied that unfortunately when a new substance appears, the first thing industry wants to know is "how much can we allow people to have" (769). "And, I have to tell you, sadly, that in the radiation field, as well as in the chemical field, in the radiation field the safe standards are set on nothing but thin air and guesswork. And, I state that flatly as a conclusion" (769–70). Gofman supported his opinion by quoting a prominent member of the Nuclear Regulatory Commission, Robert Minogue, who had written to the commissioners stating, "We should remove the term 'permissible' because it is being misused to make workers think it is safe" (774).

Q. All right, doctor. And, so when people talk, such as Mr. Paul, about a "safe standard"—

A. That is a meaningless statement. And, that is a deceptive statement if it suggests there is anything safe about it at all. . . .

Q. Suppose the evidence shows that during periods of time employees were put to work who had had no training whatsoever in understanding the hazards of radiation from plutonium?

A. I find it hard to believe that anybody would do that.

Q. Why?

A. It is unconscionable. You just shouldn't do that to humans working with something like this.

Q. What if you saw that they were trained maybe eight hours, or had eight hours training?

A. That is a cruel joke to do that. It is the same as no training at all. A very cruel joke. (780–81)

Spence was intending for Gofman and me to serve as his one-two knockout punch. Gofman certainly accomplished a first-round knockdown of Kerr-McGee.

I knew that this fight was far from over. Unfortunately, I was right in the middle of Ph.D. examinations at Georgia Tech. Spence assured me that he could get me on and off the stand in the same day. He told me I was essential to his case and that he would, if necessary, interrupt whoever else was testifying in order to put me on.

The morning of March 14 Spence introduced me to the jury and immediately read aloud highlights of my curriculum vitae, as he had done for Gofman. Calling me the founder of health physics, he described my background on the Manhattan Project and my work at Oak Ridge.

In my testimony I noted that Kerr-McGee did not even have a certified health physicist on staff and used this analogy to describe the situation:

> Sir, it is like in a hospital, you wouldn't want to operate just with interns, just students. You would want someone there with a medical degree. And in this case the equivalent of the medical degree, the certification, is to have a certified health physicist. For example, in my group at Oak Ridge National Laboratory we had 28 certified health physicists. Here they had none. Not one certified health physicist. This to me is rather deplorable. It is like, again, running a hospital on interns, students practicing on the patients, without having someone that meets the standards, the qualifications of the medical doctor. (1565–66)

Paul became agitated and requested permission to "approach the bench" after I assigned, in terms of university grading systems, "an F, a failure" to the health physics aspects of his client's facility (1568–69).

I was asked for my opinion about security of the operation as a whole if the evidence established that Kerr-McGee could not account for 40 pounds of plutonium. I noted that that would be enough to construct several nuclear weapons and would present a grave security risk for our country. Paul nearly went into orbit. He told the judge—out of hearing of the jury—that my answer was "so prejudicial as to its content, I do move for a mistrial" (1570–71).

Spence responded to Paul's motion for mistrial: "You may not like it, Mr. Paul, but that is why we are here, is because that material is of such a dangerous character that it could mean the end of all of us" (1571).

Judge Theis overruled Paul's motion—my testimony was allowed to stand. Next I had to hold up under cross-examination by the able Paul.

Paul asked whether I could identify a single reported case of cancer "which has been established to a medical certainty to have been induced by exposure to plutonium?" Instantly Spence stood and claimed the question was "tricky" as well as misleading.

Judge Theis leaned forward and asked me, "Do you understand the question?"

I replied, "Your Honor, I don't object to answering it."

The judge retorted, "I didn't assume you would, Doctor."

I stated that "I don't necessarily side with my lawyers, or anybody. That is a characteristic of a scientist" (1678–79). I then told the jury that no one can say with absolute certainty that plutonium caused a particular cancer. I testified that it is possible to establish a probability that plutonium caused a particular cancer.

Then Paul asked, "If we leave Karen Silkwood out of it, and deal with—"

Spence went ballistic. "Just a minute. I don't want to leave Karen Silkwood out of it, because that is what the case is about. We are not trying any other case ultimately . . . this is an improper and argumentative question" (1709).

Judge Theis called both counsel to the bench. In a low voice he candidly said to Paul, "You're digging a hole for yourself" (1710).

On redirect examination by Spence (which occurred after Paul had completed his cross-examination), I stated that the dangers of alpha particles were known long before Karen Silkwood died. I testified that I had examined Kerr-McGee's safety manuals and determined they fail to mention the dangers of alpha particles.

My testimony neared its conclusion:

Q. Would it be completely and utterly and wholly fair to say that by 1970, when this plant opened, four years prior to the death of Karen Silkwood, that any reasonable and responsible person in health physics knew the nature and extent, at least knew a great deal, about the extraordinary dangers of alpha particles from the standpoint of cancer?

A. In addition to the graduate doctorate course at Georgia Tech, I have an undergraduate course to teach freshmen. If I had a student on the exams that I am giving tomorrow that didn't know that, and could not elaborate on it in great detail, he would fail the course. I think anybody that has had any exposure to this field knows about the risk of alpha-emitting radionuclides.

Q. And, would they have known that fully in 1970?

A. There would have been thousands of publications that give evidence on this; yes, sir. (1730)

The jury in their deliberations relied heavily on my testimony and that of Gofman. The verdict of $10 million punitive damages and $505,000 in actual damages set a record for that type of case. Two movies, *The China Syndrome* and *Silkwood*, were based on this case.

Judge Theis, at the request of my coauthor, has prepared his independent assessment of the roles Dr. Gofman and I played in this historic controversy:[7]

> Both of these eminent scholars and scientists were subjected to comprehensive and probing questions on direct examination by plaintiff's counsel, Gerald Spence, an excellent lawyer, and to probing and exhaustive examination by the very competent staff of attorneys representing the defendant Kerr-McGee. My recollection, buttressed by the written record of the case, of which I have a copy, indicates that their testimony was unsullied in its credibility aspects. In sum, the testimony was detailed, clear and convincing.
>
> Experts are judged by the finders of fact—whether it be a jury or a judge—upon their background of experience, their learned treatises, their demeanor as a witness (in speaking authoritatively), and last and most important—on their ability to withstand exhausting and penetrating cross-examination. Cross-examination is the hallmark of our American justice system in the search for truth of the testimony of any witness—whether the case before the court be civil or criminal. A kindred quality in expert witnesses which I have noted through the years, and which impresses both the judge and jury, is their ability to be honest and open and to have the candor to discuss the pros and cons of

> their opinion—that is, acknowledging where error might lie, or other experts differ—and finally, to explain why their opinion is the sound one.
>
> Reviewing in my own mind the many excellent experts in areas of human endeavor that I have heard during my twenty-four years on the federal trial bench, I believe these two gentlemen are among the most impressive, articulate, honest, and knowledgeable expert witnesses in their respective fields that I have had the privilege to hear and observe.

The real victory in the *Silkwood* case is that it brought to the forefront one of the worst fears of the nuclear industry: educating the public that there is no such thing as a "safe dose" of radiation.

A few years later I was asked to testify in a case that challenged the credibility of my former employer, the U.S. government. A blockbuster case brought by 1,192 plaintiffs, *Allen v. United States* stretched our nation's judicial capacity to its limits.[8] No jury determined the outcome because under the Federal Torts Claims Act, citizens who sue the federal government are denied a trial by jury. Instead, Judge Bruce Jenkins of the U.S. District Court for the District of Utah served as the sole trier of fact (fig. 52). Unlike some other judges, Jenkins took it upon himself to gain a solid grasp of nuclear physics before rendering his decision.

The *Allen* case involved claims against the United States to recover damages for cancer and leukemia resulting from desert weapons tests of atomic devices.

Although *Allen* did not involve nearly as much money as *Silkwood*, a different issue was at stake: the integrity of our government.[9] The Utah judge would determine whether the United States intentionally concealed information at high levels of responsibility. Once again Gofman and I agreed to testify for the underdogs.

On the morning of October 7, 1982, Dale Haralson, one of plaintiffs' counsel, called me to the witness stand. Haralson had spent countless hours preparing the case, knowing full well he would never recover his regular hourly rates, even if he won.

He commenced by covering my background in physics and health physics, both from an educational standpoint and at Oak Ridge National Laboratory.

My principal task in this case was to establish what the government knew or should have known about the risks of exposure to radiation and what safety measures should have been followed regarding the Nevada tests.

Fig. 52. Judge Bruce S. Jenkins, who served as the sole trier of fact in Allen v. United States, *in which I testified as a key expert witness. (Photo courtesy of Judge Jenkins)*

Judge Jenkins listened intently as I explained how fission occurs within millionths of a second with a U-235 atom.

I told the court that alpha particles present little damage outside the body but that "once in your body, they are the worst particles you can conceive" because they concentrate in radiosensitive tissues where cancers originate (2761–66).[10]

Then Haralson asked, "Was this something that was known in the field of physics in the mid-40s?" I responded, "It's information that I learned back . . . in say, 1925 to 1934" (2767).

When asked about cell damage from gamma rays, I explained that these rays penetrate the entire body. When they strike a cell one of four things happens: (1) the gamma wave does no damage; (2) it causes the entire cell to die; (3) the radiation actually breaks a chromosome or knocks out a piece, and the cell manages to repair itself; or (4) the gamma radiation damages the nucleus of the cell, but the cell survives in its damaged form and reproduces itself in its aberrant characteristics. "So this cell survives today it's one cell, day after tomorrow there are two cells. A week later there's a dozen cells, then five years later there are ten thousand. Eventually now this is a big enough colony of cells that it's diagnosed as a malignancy" (2774–75).

Judge Jenkins appeared eager to learn as much about ionizing radiation as possible. The courtroom was becoming the classroom, and I sensed that he wanted me to serve as his tutor.

I stated that variations in sensitivity play an important role in radiation-induced cancers: "First of all, I'm not the same person I was yesterday, and you're not the same as you were yesterday. We're all changing constantly. . . . The chemistry of our system is changing constantly, and we react differently to the same stimuli depending on when and under what circumstances" (2783).

I testified that there is great variation among individuals and that children are more radiosensitive than adults. "For some types of cancer [such as thyroid], the female is more susceptible" (2783).

I made clear that a responsible health physicist would have realized that exploding an atom bomb causes millions of curies of radioactivity to be released and that a large number of monitoring and testing measures would be mandatory.

> I would have set up a large number of environmental monitors as I did at Oak Ridge [which had orders of magnitude less exposure] out to great distances. . . .
>
> I would have collected urine and fecal samples from people at the site and off-site in the nearby cities as we did in our program. I would have collected blood samples.
>
> And one thing that we learned early at Oak Ridge and it was impressed on us, which I was asked to go to Windscale and discuss with them after their serious reactor accident, that light airplanes can be a lifesaver in such cases. . . . So, at Oak Ridge we had instruments that we could fit into boxes and rush to the little airport there. And in our drills, we were actually able to get airborne within a half hour. (2822–23)

I noted that the use of light airplanes with monitoring equipment revealed hot spots never anticipated from our ground testing.

We discussed a document dated September 16, 1953, prepared by the federal government at its Los Alamos laboratory. The document stated that the U.S. government had determined to forego collecting milk samples from cows in off-site locations since such activity could "alarm an already worried community" (2829).

Haralson asked, "Would that be consistent with good health physics, Dr. Morgan?" I responded, "No, I would hesitate to say what I think it involves. But it certainly would not be in conformance with the spirit in which health physics is hopefully practiced" (2830).

Judge Jenkins listened intently as I explained our misconception in the early period that if one did not receive a dose above a certain threshold, the body repair mechanism would repair any cell damage.

I told how scientists later came to accept the linear hypothesis, that all radiation is harmful. Haralson asked if I myself subscribed to the linear approach. I replied that I had "gone one step further" and possessed "good evidence" to support the "super linear hypothesis," which is that "with very low doses, you get more cancers per rem than you do at high doses. I didn't say that low doses produce more cancers than high. I said you get more cancers per rem at low doses than at high doses" (2847–48).

I informed the judge that I had traveled to the Nevada test site several times during the early 1950s: "I wasn't able to find any record of measurements that were made off-site in reference to internal dose. . . . [I] saw no evidence of what I would consider an environmental monitoring program for the public. . . . I saw no data from people off-site as far as total body or internal exposure" (2854–55).

I testified that one means of indirectly monitoring internal dose to off-site humans is to analyze the food chain in the area. By examining the liver and thyroid of field mice and the thyroid of cows and hogs in the off-site area, a health physicist would have a good idea of the amount and kinds of radionuclides deposited in the thyroids and other body organs of children in the area (2860–61). I stated that "we discussed this in our seminars, that tissues from hogs, the gonads, the thyroids, slices of liver would give us an excellent idea of what man might be taking in because their food intake is very similar to that of man" (2861).

Haralson asked me to explain the significance of sheep, cattle, and horses in fallout areas that "developed lesions on their mouths, . . . epilation or loss of wool and hair, . . . scarring and ulcerations and sores" (2814–15). I stated that those symptoms were indicative of exposures in local doses in the thousands of rads (2814–15).

Haralson presented a letter written in June 1953 by the project veterinarian to Los Alamos stating that sheep losses in the area resulted from radiation exposure and recommending that "a detailed study must be made" and that all the animals should be moved a greater distance from the test fallout area (2865).

When asked what information I would have shared with the public about radioactive fallout, I stated, "The people deserve to know what the risk is but they should make the decision how much radiation they want to take or are willing to take and there shouldn't be coverups . . . I've always felt that information about the risks should be made freely available to the public, and I think the public can be understanding" (2869).

Upset with my answer, Mr. Chavey, counsel for the United States, jumped up and asked that it be stricken from the record. Judge Jenkins looked at him and stated, "He's merely suggesting people be told the truth, as I understand his answer . . . I'll let the answer stand" (2870).

The court permitted me to explain the significance of excess leukemias in a population group. I concluded my direct examination by pointing out the necessity of meeting with medical groups in off-site areas to alert them about the type of malignancies to expect (2881). By the time I had finished, it was 7:10 P.M. I was exhausted and would be required to return in one week for cross-examination.

The delay benefited the government by allowing it an entire week to prepare its case.

The government designated Henry A. Gill, acting assistant general counsel, to cross-examine me. Apparently Gill thought he could trap me by asking, "All the answers [on the dangers of radiation] aren't in, is that correct?" I replied, "I'd hate to live in a world where we had all the answers. That's one of the thrilling parts of existence, that every door we open we see a dozen more that we would like to enter" (3218–19). I glanced over at Judge Jenkins. Although he wore a poker face, I had a gut feeling that he trusted me.

That afternoon Gill ended his cross-examination by reading ad nauseam, ad infinitum from the *Federal Register*. This exercise in futility appeared to annoy Judge Jenkins, whose mind worked orders of magnitude faster than Gill spoke.

On redirect examination by Haralson, I made clear that the linear dose concept was accepted by many scientists before the weapons testing in the early 1950s. I also stated that hundreds of publications existed during this period showing the adverse effects resulting from relatively low doses of radiation to animals.

Gill objected again, but the judge permitted me to express my opinion that if animal studies in the 1950s had shown excess cancers at low doses, the government should have assumed that the same dose would have the same effect on humans (3318–19).

Time was running short, since my flight was due to leave in one hour. I concluded by telling the court that sufficient information existed by the time of the weapons tests for the government to know the carcinogenic effect of low-level radiation on human beings (3328). I explained why the government could not in good faith claim ignorance: "With the meetings of the Directors of Health Physics and Biology and Medicine which we held at the site and all of the other Atomic Energy installations in my discussions with other directors, I felt that they, too, appreciated the fact that there are serious long-term risks as a result of the fallout" (3329).

I knew that the most important person in the courtroom appreciated my testimony when Judge Jenkins excused me from his court, saying, "We appreciate your help and I have enjoyed your class" (3331).

As I rushed to catch my flight, I had mixed feelings about the case. On the one hand, it was obvious the court trusted and understood me. That made me feel good. On the other hand, the fact that our government was denying any knowledge that low-level radiation presented serious health consequences appalled me.

The next morning Ralph Hunsaker, one of plaintiffs' counsel, called Gofman to the witness stand. Gofman explained the human cell system and its information library just as he had done in the *Silkwood* case.

Gofman stated that an atomic bomb blast produces alpha, beta, and gamma radiation in the fallout over the countryside (3470–71).

When asked whether there is such a thing as a "safe dose" of radiation, Gofman replied: "There has never been in the history of science any evidence that there's a safe dose of radiation. . . . There is no threshold. There has never been a shred of evidence for it, and you can expect cancer or leukemia or chromosome injury harm all the way down to the lowest conceivable dose" (3473–74).

Gofman confirmed that radiation doses of 50–100 rads produce nausea, vomiting, skin burns, hair loss, and depression of sperm count. Then Hunsaker asked whether, if a person has these symptoms after a fallout cloud has passed over, the cause of the symptoms is radiation. Gofman replied: "There is a rule in science that's called Occam's razor, and any good scientist will always follow it and the rule says: don't seek multiple causes when you have one good one" (3476).

Gofman verified a report that counties nearest the Nevada test site experienced a 344 percent increase in leukemia (3513). Continuing his testimony, he explained various cancers caused by radiation and the dose necessary to produce each such cancer.

His direct examination concluded by disclosing the following discussion with Charles Dunham, former director of the AEC Biology and Medicine Department. In 1963 Dunham, Gofman's Washington boss, directed Gofman to attend a meeting but refused to tell him in advance the purpose of the meeting.

> So I went into Washington and present there were myself and five other people. I think four out of five at least were scientists, and none of us knew what the meeting was about until Dr. Dunham and Dr. Gordon came into the room and explained to us. . . . Dr. Dunham explained to the group the reason why they had been called together. He said there was a man in the biology and medical division by the name of Dr. Harold Knapp who had been doing analysis of the dose to the thyroid gland and that he had found that the true dose to the thyroid gland was many, many times higher than has been reported by people associated with the Atomic Energy Commission. I don't remember the exact number of times more, but it was in the ballpark of a hundred times more. And the group asked what is it that you are telling us this for? . . . He said, if Dr. Knapp publishes his findings, the public will know that they haven't been told the truth up to now, and my request to this committee is to see whether you can convince Dr. Knapp not to publish his findings. (3604–5)

Gofman and his colleagues refused to buckle under and insisted that "the appropriate thing for the Atomic Energy Commission to do would be to have Dr. Knapp publish his findings" (3606).[11]

Judge Jenkins heard testimony from ninety-eight witnesses whose testimony in the trial transcript exceeded seven thousand pages. He reviewed more than fifty-four thousand pages of written material in the form of trial exhibits. His 225-page opinion exhibited a rare judicial appreciation of nuclear physics, health physics, and medicine.

The judge's conclusions represented a resounding victory for truth:

> Although some scientists and commentators have at times suggested the presence of a "threshold" dose, the predominant philosophical approaches to radiation protection have carefully eschewed such a view, and the overwhelming weight of currently available scientific evidence supports the view that at *any* exposure level, ionizing radiation causes *some* degree of biological damage and creates *some* long-term risk of cancer and leukemia in those persons who are exposed.[12]

Judge Jenkins later shared with my coauthor his opinion of the contributions Gofman and I had made to the case:

> I respond with pleasure to your inquiry concerning my experience with Dr. Karl Z. Morgan and Dr. John W. Gofman during the course of a case heard by me.
>
> The case was *Allen v. U.S.* and my opinion is found at 588 F. Supp. 247 (1984).
>
> Doctors Morgan and Gofman appeared as expert witnesses. Both were called by plaintiffs. In addition to expertise, each provided historic data.
>
> It appeared to me at trial that each was a scientist in the best of senses. Each had high regard for truth. Each was gifted, knowledgeable, articulate.
>
> Equally impressive to me as their intellectual capacity and depth of knowledge, was what appeared to me to be their selfless motivation—a genuine concern for people. I admired that. I still do.
>
> In crafting my opinion, I relied greatly on their testimony, their knowledge, their writings, their expertise. The opinion makes reference to them.
>
> I am happy to give you my brief impression of what to me are two persons of historic dimension and great merit.[13]

The victory in the *Allen* case established that the U.S. government knew about the hazards of fallout and "failed to adequately and continuously inform individuals and communities near the test site of well-known and inexpensive methods to prevent, minimize or mitigate the known or foreseeable long-range biological consequences of exposure to radioactive fallout, and that such failure was negligent."[14]

Until this case, I did not fully appreciate how difficult it is to recover a judgment against the United States under the Federal Torts Claims Act (FTCA). When fighting the federal government in court, average citizens start the case with two strikes against them. First, they must overcome an adversary with unlimited resources. Second, they face a stacked deck in terms of the Federal Torts Claims Act, which is filled with loopholes designed to protect the government.

The *Allen* decision was reversed by the Tenth Circuit Court of Appeals because of a provision in the FTCA called the "discretionary function exception." This exception to government liability is so broad that it encompasses "social, economic, and political decision making" of the United States.[15] Judge McKay of the circuit court wrote that "while we have great sympathy for the individual cancer victims who have borne alone the costs of the AEC's choices, their plight is a matter for Congress.

Only Congress has the constitutional power to decide whether all costs of government activity will ... continue to be unfairly apportioned, as in this case."[16]

The FTCA's philosophy of "the king can do no wrong" does not comport with having a government "of the people, by the people, and for the people." Sovereign immunity can never be accepted where injured citizens have demonstrated that the government's negligence has caused their maladies.

CHAPTER TEN

A Time for Reflection and Resolution

There is a time for everything, and a season for every activity under heaven: a time to be born and a time to die.

ECCLESIASTES 3:1–2

Today, at age ninety-one, I am close to completing the circle of life. Thanks to God, my mind functions almost as well as ever. Now is the time for me not only to reflect on my principles and my past but also to look ahead and urge those who will survive me to maintain the resolve necessary to overcome two of the greatest challenges of the nuclear era.

I began life on planet earth in a world of dreams. I believed anything was possible with prayer and hard work. I still do. Now is the time for me to reflect upon my faith—the driving force in my life—and to examine the resolve that I believe will be required to avert or limit nuclear catastrophes in the years ahead. This chapter completes the circle from where I started to where I will soon end.

REFLECTIONS ON A SCIENTIST'S FAITH

My life began with plans and ambitions to be a Lutheran missionary. Instead, I became a physicist and, more specifically, a health physicist. I would like to think that this change was the will and act of the God in whom I trust. I believe the change made me a stronger Christian and supported me with a more meaningful belief and trust in God.

As a scientist I believe that God began this universe, consisting of billions of galaxies each with billions of suns, some 10 to 15 billion years ago with what we call the Big Bang. I believe that time, space, and matter did not exist before then. At the instant of the Big Bang, time and space flashed into existence. Shortly after this, I believe, matter in various forms began to come into existence.

I believe that human beings have some form of existence after death, but not as a human body through which food, water, and air must pass as excretions. I believe that there is a heaven, but not a city floating in the sky. And like my father, I believe that there is a hell, the home of the devil.

As a human I see, in the words of the apostle Paul, "but a poor reflection as in a mirror," but after my inevitable death I will see clearly "face to face" in heaven, see as I am seen and understand what now in this life are mysteries. I have been blessed here with a few answers and understandings, but each brings more mysteries. I believe that for things to exist or for things to happen, there must be a force to make them exist and happen, and there must be an awareness and a cause of things that exist and of things that happen, and for me this is the mystery and embodiment of God. I believe that this God is omniscient and omnipotent. I believe that He knows everything that happens, has happened, or will happen. I believe that He knows every thought and hears every wish and every prayer of every human being, though I don't understand how He knows my every thought. I believe that the closest we can come to knowing who God is, is the notion that God is love. Most of all, I don't understand why He loves me and you.

As a scientist and as a Christian, there are many things I don't understand. But all these mysteries don't make me less a scientist or less a Christian. For example, as a scientist I don't understand how there could be a beginning of time, space, and matter, and I wonder if there is an end. Is there enough mass in the reaches of space that its rate of expansion is slowly declining? If so, will the universe eventually start contracting, leading in a trillion years to another Big Bang?

I realize that the possible time of existence of any life on our earth is only a moment in the time of existence of our solar system. Thus, as God took on human form in Christ 2000 years ago, this was an extremely special and important event not just for earth but for the universe and for all time and all space. It was His gift and promise and proof that those who accept Him will live eternally in some advanced and most desirable form that has no end and no sadness. Why does life exist only on earth for this moment of time? Does God have some great plan for the use of the billions of galaxies and the billions of suns in each and of the planets revolving about some of them? Or has the glass through which I see become too darkened?

I and all surviving forms of life learn by experience, and we also profit by mistakes. I hope and pray that this book will lead to greater respect for both the good and the evil of atomic energy and that through love of one another, human beings will choose safe use of atomic energy for peaceful purposes. Just as there is a time for everything, I believe that God has granted me time to complete this book before I complete the circle.

NEED FOR FOCUSED RESOLVE

Exercise of First Amendment Rights

History teaches us that there is no more effective means of bringing about a change in the conduct of government or industry than the persistent efforts of citizens. Nothing can replace persistence. Talent and genius will not, nor will education. Persistence and determination are powerful.

Big industry has also realized the omnipotence of a persistent citizenry. Unfortunately, strategic lawsuits against public participation (SLAPP) suits are occurring with increased frequency in this country.[1]

More than twenty-five years ago, Nancy Hsu Fleming immigrated to the United States. Upon becoming a citizen, she never imagined "she could get hauled into court for exercising her constitutional rights."[2] Fleming, a resident of North Kingstown, Rhode Island, thought she was exercising her First Amendment right to petition the government when she wrote to the Rhode Island Department of Environmental Management. In her letter, which was copied to the governor, Rhode Island's congressional delegation, and several EPA officials, Fleming expressed concerns about possible groundwater pollution of the local drinking water supply caused from a landfill. The landfill owners responded with a lawsuit for defamation and tortious interference with prospective business clients. Although Fleming ultimately prevailed, the landfill owners "got what they wanted" by essentially silencing her for four years.[3]

Justice Nicholas Colabella of the New York Supreme Court for Westchester County addressed the issue directly: "SLAPP suits function by forcing a target into the judicial arena where the SLAPP filer foists upon the target the expenses of a defense. The ripple effects of such suits in our society are enormous." Judge Colabella appropriately warned that "persons who have been outspoken on issues of public importance targeted in such suits or who have witnessed such suits will often choose in the future to stay silent. Short of a gun to the head, a greater threat to the First Amendment expression can scarcely be imagined."[4]

In view of growing concern over SLAPP suits, some state legislatures have passed laws "to counter their chilling effect."[5] Most anti-SLAPP statutes are designed to require an early court analysis of the merits of a SLAPP-type claim. Once a defendant requests such a review by the court, it becomes the burden of the company or individual filing the suit to convince the court it is meritorious. Otherwise, it will be dismissed early in the proceedings.

No society that severely restricts freedom of speech will ultimately survive. It is one thing to know that government or industry needs to make a change. It is quite another to stand up and insist that the change be made. The world community of informed citizens simply must make government and the global nuclear-industrial complex accountable in two key areas, nuclear waste disposal and nuclear weapons arsenals.

Nuclear Waste Disposal

At each of the 109 nuclear power plants in the United States, every three to four years the "spent" (partially depleted) fuel assemblies in the reactors must be removed and replaced. These radioactive spent assemblies of rods contain a high concentration of fission products and transuranium radionuclides, such as plutonium-238, -239, -240, -241; americium-241, -243; and curium-245, -246, -247. These are considered among the most dangerous materials known. By 1989 seventy of these plants had accumulated some 20,000 metric tons of spent nuclear fuel. At the end of the license period of all existing nuclear power plants, the amount of spent nuclear fuel is expected to exceed 80,000 metric tons.[6]

The plutonium can be removed from these fuel elements and mixed with uranium to serve as fuel for nuclear power plants. In this case, the U-238 enrichment for power reactors would be from Pu-239 instead of from U-235. Although Japan and perhaps Germany and France may move in this direction, I do not believe this to be a good choice. The best move for the present time is to let the fuel assemblies cool in the power reactor "swimming pools" for at least three years to permit radioactive decay of some of the short-lived fission products and then ship them to a temporary and retrievable storage facility. Then it would be up to a future society to determine the best next step. It is my hope that fifty or a hundred years from now society will have advanced and will have arrived at the best long-term solution.

"Burning" plutonium is the only way to destroy it and thereby prevent its use in nuclear weapons. The storage facility would have to provide adequate cooling of the fuel assemblies and prevent their corrosion. Most important, the facility would be required to maintain adequate security and would be subject to supervision of the International Atomic Energy Agency (IAEA) at all times. Perhaps future society will decide it is not safe or feasible to build specially designed nuclear power plants to "burn" plutonium-spiked uranium. At that time the assemblies could be stored in salt formations, as described in chapter 8.

Nuclear waste in the former Soviet Union has reached alarming proportions. For example, in the first two years of operations of the Chelyabinsk-40 complex, high-level nuclear waste was simply dumped into the Techa River, causing radionuclides to enter the Arctic Ocean. To alleviate this contamination problem, plant wastes are now discharged into Lake Karachai, which does not have an outlet. The lake has accumulated 120 million curies of long-lived radionuclides and is so "hot" that a person standing on the shore will receive a lethal dose (about 600 R) in an hour. The Techa River itself is cordoned off with a wire fence. People are prohibited from fishing in it or even cutting hay nearby. "There are 400 million cubic meters of radioactive water in open reservoirs along the river. Fish in one reservoir are reported to be '100 times more radioactive than normal.'"[7]

Nuclear Arsenals

Nuclear arsenals of a dozen countries present our greatest challenge. The most serious threat is the result of lax nuclear security in the former Soviet Union. Physicist Frank Von Hipple, while a White House representative in 1994, visited several of Russia's nuclear facilities.[8] Taken to a warehouse in the Ural Mountains, he observed twelve thousand steel canisters, each containing more than 5 pounds of weapons-usable powdered plutonium oxide.[9]

Just two of the canisters contained enough plutonium to construct the bomb dropped on Nagasaki. The warehouse was located in a secret nuclear city the Soviets named Chellubinsk 65. Surrounded by a fence and guarded by thousands of Russian troops, the city was reasonably secure from an outside attack. However, security measures to prevent "an inside job" were virtually nonexistent: no video surveillance, no radiation detectors, no motion sensors or alarms, no identification cards, no code numbers, and no sensors connected to each canister of plutonium (to alert authorities if someone tried to move a canister). This facility seemed a perfect target for terrorists. According to Von Hipple, "there are virtually no safeguards against this type of inside job, the kinds of safeguards we have in U.S. facilities."

Deplorable security conditions are the rule at most of the nearly fifty Russian sites that house uranium and plutonium weapons. Graham Alison of Harvard University discovered that as of April 1996, in at least 80 percent of the facilities where these weapons are stored, "there is no electronic perimeter system, not even the control system you would have when going into the airport."

Even the director of the CIA publicly admits the severity of the problem: "Due to severe research shortages the Russian nuclear weapons complex is deteriorating and it continues to be a serious threat for diversion of nuclear technology and materials to other proliferating countries in the world."

Ironically, the problem is getting worse as a result of the Strategic Arms Reduction Agreement (START), which requires Russia to dismantle over two thousand nuclear warheads a year. Since Russia is unable to pay all its nuclear scientists a living wage, they become tempted to enter the nuclear black market. A flyer circulating in the Middle East advertised that a Hong Kong company could make available hundreds of former Soviet nuclear weapons scientists who would purportedly work for other countries for reasonable pay.

Russia's entire nuclear stockpile comprises more than 1000 tons of weapons-grade uranium and more than 200 tons of plutonium. As these warheads become dismantled, they are, in many cases, sent to insecure facilities. Senator Sam Nunn firmly believes the lax Russian nuclear security measures present "the number one national security challenge this country faces today and will be the number one security challenge we face for years to come."

The chief of the FBI's Domestic Terrorism Division admits that "the growing strength of international organized crime, especially within the former Soviet Union republics, has raised the specter of an underground market for these . . . weapons of mass destruction to terrorist groups and to rogue nations."

Perhaps more disturbing is our government's admission that it is "very poorly equipped to protect our infrastructure from a terrorist attack based on nuclear, chemical, or biological weapons."

One way to solve the Russian problem at least in part would be for the United States or a consortium of peacekeeping nations to buy Russia's nuclear arsenal and then secure it. Graham Alison, who has studied this problem extensively, makes a good point when he says: "Suppose I can buy the weapons and weapons equivalent for $1 million each . . . for $30 billion I can buy 30,000 nuclear weapons or nuclear weapons equivalent. If I ask myself, 'Would I like for us to buy them and take them or to leave them out there for somebody else?' You bet—$30 billion would be the best buy we ever made in our defense budget."

Time is not on our side. The international community of responsible nations must move quickly.

WHICH WILL IT BE: THE BLESSING OR THE CURSE?

Today, more than half a century since humankind unleashed the energy of the genie for good or evil, I ask, "Which will it be?" Is it the beginning of the end or the start of a period of great intellectual advancement and miraculous achievement?

The genie cries aloud for wisdom and understanding to prevail. If we can use the genie properly and minimize the risk we have created by misuse, great challenges and opportunities exist for a happier life for all humankind and the opportunity to fulfill human dreams and ambitions. Should we fail in this quest, the adverse consequences will reverberate through future generations. I believe the future will be either the last frontier for mankind or a new period of cooperation and advancement for the world.

APPENDIX ONE

Albert Einstein to President Roosevelt, August 2, 1939

Albert Einstein
Old Grove Rd.
Nassau Point
Peconic, Long Island

August 2nd, 1939

F.D. Roosevelt,
President of the United States,
White House
Washington, D.C.

Sir:

Some recent work by E.Fermi and L. Szilard, which has been communicated to me in manuscript, leads me to expect that the element uranium may be turned into a new and important source of energy in the immediate future. Certain aspects of the situation which has arisen seem to call for watchfulness and, if necessary, quick action on the part of the Administration. I believe therefore that it is my duty to bring to your attention the following facts and recommendations:

In the course of the last four months it has been made probable—through the work of Joliot in France as well as Fermi and Szilard in America—that it may become possible to set up a nuclear chain reaction in a large mass of uranium, by which vast amounts of power and large quantities of new radium-like elements would be generated. Now it appears almost certain that this could be achieved in the immediate future.

This new phenomenon would also lead to the construction of bombs, and it is conceivable—though much less certain—that extremely powerful bombs of a new type may thus be constructed. A single bomb of this type, carried by boat and exploded in a port, might very well destroy the whole port together with some of the surrounding territory. However, such bombs might very well prove to be too heavy for transportation by air.

The United States has only very poor ores of uranium in moderate quantities. There is some good ore in Canada and the former Czechoslovakia, while the most important source of uranium is Belgian Congo.

In view of this situation you may think it desirable to have some permanent contact maintained between the Administration and the group of physicists working on chain reactions in America. One possible way of achieving this might be for you to entrust with this task a person who has your confidence and who could perhaps serve in an inofficial capacity. His task might comprise the following:

a) to approach Government Departments, keep them informed of the further development, and put forward recommendations for Government action, giving particular attention to the problem of securing a supply of uranium ore for the United States;

b) to speed up the experimental work, which is at present being carried on within the limits of the budgets of University laboratories, by providing funds, if such funds be required, through his contacts with private persons who are willing to make contributions for this cause, and perhaps also by obtaining the co-operation of industrial laboratories which have the necessary equipment.

I understand that Germany has actually stopped the sale of uranium from the Czechoslovakian mines which she has taken over. That she should have taken such early action might perhaps be understood on the ground that the son of the German Under-Secretary of State, von Weizsäcker, is attached to the Kaiser-Wilhelm-Institut in Berlin where some of the American work on uranium is now being repeated.

Yours very truly,
Albert Einstein [signed]

APPENDIX TWO

A Preliminary Report on the Low-Draft Fly Swat, April 1, 1946

This document consists of 1 page and no figures. Copy 7 of 9 copies, Series A

April 1, 1946

1. J. E. Wirth
2. M. D. Whitaker
3. R. C. Thumser
4. L. W. Nordheim
5. M. C. Leverett
6. J. R. Coe
7. K. Z. Morgan
8. Central File
9. Readers File

To: J. E. Wirth
From: K. Z. Morgan
In Re: A PRELIMINARY REPORT ON THE LOW-DRAFT FLY SWAT

This instrument is one of the most useful instruments in the laboratories, counting rooms and general operating areas of Clinton Laboratories during the summer months. Since other pre-war instruments such as G. M. tubes, scaling circuits, electrometers, alpha counters, etc., must be written up and described in detail in classified secret reports, I believe that this instrument should be described in similar detail. This will serve only as a preliminary report.

There are a number of types of Fly Swats in general use at Clinton Laboratories. The Low-Draft is in most common use and is the subject of this report.

This Low-Draft Fly Swat is rather conventional in that is has a wooden handle about 12″ long. The portion of the swat that comes in contact with the fly is made of rubber covered cloth. It is 3½″ wide and the useful portion for contacting the fly is about 3½″ long. This rubber covered cloth contains 80 holes 1/8″ in diameter. The purpose of these holes is to reduce the compression wave when swatting a fly. Otherwise the air concussion would help the fly on its take-off and would give him additional warning of imminent danger. It would be desirable to have a good supply of these fly swats by early spring because considerable time is spent during the summer months by highly paid personnel contesting with the flies on the right to operate the laboratories.

As a hint for better use of a fly swat one might remember that most flies take off backwards and one should strike at a point slightly behind the fly. As a final remark, I would like to suggest that more janitors be employed during the summer months to assist in removing the fly carcasses from the floors. They represent a physical hazard in that one might slip on them and fall.

Karl Z. Morgan [signed]

This document contains information affecting the national defense of the United States within the meaning of the Espionage Act, USC 50: 31 and 32. Its transmission or the revelation of its contents in any manner to an unauthorized person is prohibited by law.

APPENDIX THREE

Aide-Memoire of Conversation between President Roosevelt and Prime Minister Churchill, September 18, 1944

10 Downing Street, Whitehall.

TOP SECRET

TUBE ALLOYS

Aide-memoire of conversation between the President and the Prime Minister at Hyde Park, September 18, 1944.

1. The suggestion that the world should be informed regarding Tube Alloys, with a view to an international agreement regarding its control and use, is not accepted. The matter should continue to be regarded as of the utmost secrecy; but when a "bomb" is finally available, it might perhaps, after mature consideration, be used against the Japanese, who should be warned that this bombardment will be repeated until they surrender.

2. Full collaboration between the United States and the British Government in developing Tube Alloys for military and commercial purposes should continue after the defeat of Japan unless and until terminated by joint agreement.

3. Enquiries should be made regarding the activities of Professor Bohr and steps taken to ensure that he is responsible for no leakage of information, particularly to the Russians.

FDR WC [signed]
18–9

APPENDIX FOUR

Albert Einstein to President Roosevelt, March 25, 1945

112 Mercer Street
Princeton, New Jersey
March 25, 1945

The Honorable Franklin Delano Roosevelt
The President of the United States
The White House
Washington, D.C.

Sir:

I am writing you to introduce Dr. L. Szilard who proposes to submit to you certain considerations and recommendations. Unusual circumstances which I shall describe further below induce me to take this action in spite of the fact that I do not know the substance of the considerations and recommendations which Dr. Szilard proposes to submit to you.

In the summer of 1939 Dr. Szilard put before me his views concerning the potential importance of uranium for national defense. He was greatly disturbed by the potentialities involved and anxious that the United States Government be advised of them as soon as possible. Dr. Szilard, who is one of the discoverers of the neutron emission of uranium on which all present work on uranium is based, described to me a specific system which he devised and which he thought would make it possible to set up a chain reaction in unseparated uranium in the immediate future. Having known him for over twenty years both from his scientific work and personally, I have much confidence in his judgment and it was on the basis of his judgment as well as my own that I took the liberty to approach you in connection with this subject. You responded to my letter dated August 2, 1939 by the appointment of a committee under the chairmanship of Dr. Briggs and thus started the Government's activity in this field.

The terms of secrecy under which Dr. Szilard is working at present do not permit him to give me information about his work; however, I understand that he now is greatly concerned about the lack of adequate contact between scientists who are doing this work and those members of your Cabinet who are responsible for formulating policy. In the circumstances I consider it my duty to give Dr. Szilard this introduction and I wish to express the hope that you will be able to give his presentation of the case your personal attention.

Very truly yours,
A. Einstein [signed]

APPENDIX FIVE

"Urgent" Message "To the President from the Secretary of War," Memo 41011, July 30, 1945, and President Truman's Reply

(reproduced with permission from the Harry S. Truman Library)

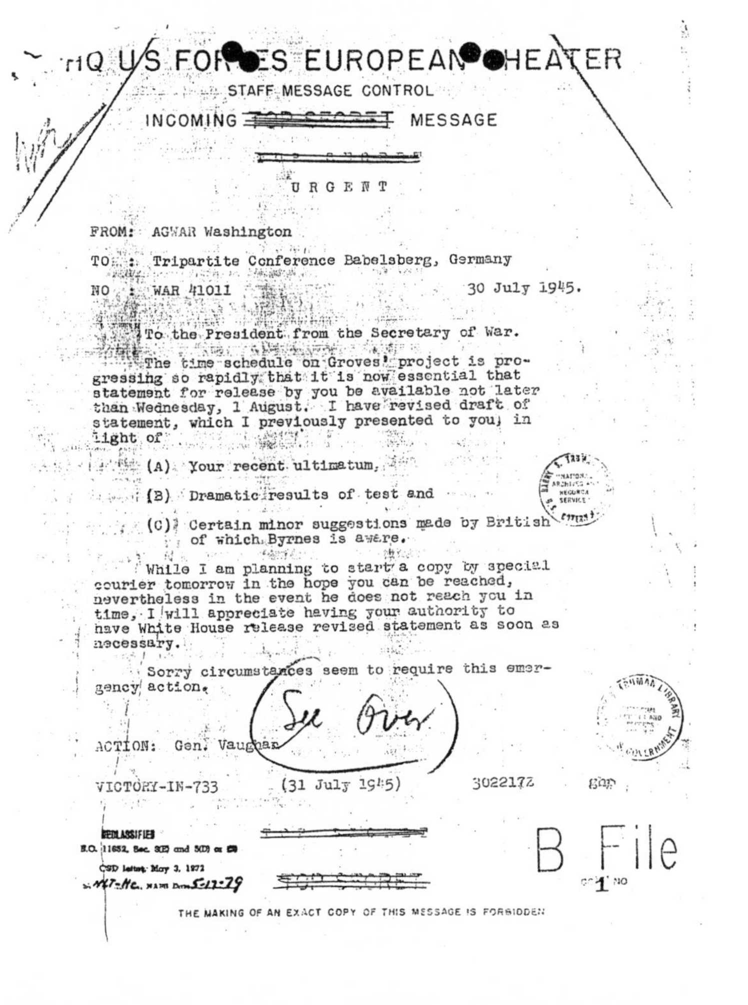

HQ US FORCES EUROPEAN THEATER

STAFF MESSAGE CONTROL

INCOMING ~~TOP SECRET~~ MESSAGE

U R G E N T

FROM: AGWAR Washington

TO : Tripartite Conference Babelsberg, Germany

NO : WAR 41011 30 July 1945.

To the President from the Secretary of War.

The time schedule on Groves' project is progressing so rapidly that it is now essential that statement for release by you be available not later than Wednesday, 1 August. I have revised draft of statement, which I previously presented to you, in light of

(A) Your recent ultimatum,

(B) Dramatic results of test and

(C) Certain minor suggestions made by British of which Byrnes is aware.

While I am planning to start a copy by special courier tomorrow in the hope you can be reached, nevertheless in the event he does not reach you in time, I will appreciate having your authority to have White House release revised statement as soon as necessary.

Sorry circumstances seem to require this emergency action.

See Over

ACTION: Gen. Vaughan

VICTORY-IN-733 (31 July 1945) 302217Z

DECLASSIFIED
E.O. 11652, Sec. 3(E) and 5(D) or (E)
OSD letter, May 3, 1972
By NLT-HC, NARS Date 5-17-79

B File

THE MAKING OF AN EXACT COPY OF THIS MESSAGE IS FORBIDDEN

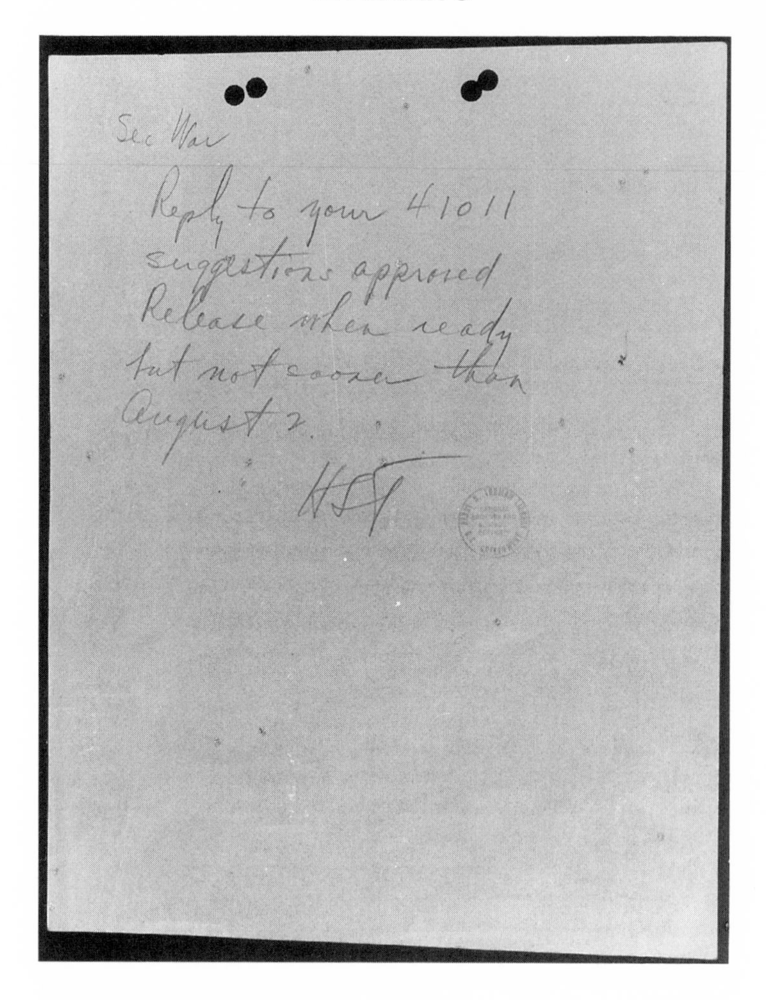

Sec War

Reply to your 41011
suggestions approved
Release when ready
but not sooner than
August 2.

HST

APPENDIX SIX

Honorable Frank G. Theis to Whom It May Concern, December 10, 1990

United States District Court
District of Kansas
Chambers of Frank G. Theis, Senior Judge
Chief Judge Emeritus
414 United States Courthouse
Wichita, Kansas 67202
December 10, 1990

To Whom it May Concern:

I have been asked to write an independent appraisal of Dr. John W. Gofman and Dr. Karl Z. Morgan in their capacities as expert witnesses. Normally, as a sitting Senior United States District Judge, I would be most reluctant to express an opinion on the merits or demerits of any witness who testified in a trial before me, except as was necessary in an opinion of the judge after a court trial. However, in view of my judicial status as a senior judge, and other factors, I consider it extremely unlikely that I shall ever again be called upon to sit on a court trial in which either of them might be a witness, and in which case I would recuse. I am therefore pleased to give my personal opinion of them as expert witnesses.

As a further prefatory statement, let me say that I have sat on hundreds of complex and protracted cases in which some of the world's foremost experts in different branches of science, art, literature and education, have testified before me. The development of nuclear energy in the Manhattan Project and outer-space exploration have been two of the greatest achievements of our millennium, and I have heard expert testimony involving both areas.

In 1979 I had the privilege of being the sitting judge on the trial of *Silkwood v. Kerr-McGee Corporation*, which was the first case involving tort liability for operation of a nuclear facility. The trial was conducted in Oklahoma City, was quite adversarial, and lasted some thirteen weeks. The *Silkwood* case was a jury trial in which the jury determined the facts. In that case I authored several major opinions the principal one involving the factual background, the trial contentions of the parties, and the law of the case, which can be found in 485 F.Supp. 566 (1979). It was a unique, fascinating and well-tried case.

Both Dr. Gofman and Dr. Morgan testified as expert witnesses on behalf of the plaintiff. Dr. Gofman was one of the foremost authorities and pioneers in the discovery of Uranium 233 and its fissionability. Because of Dr. Gofman's work with Uranium 233, the world now has access to a major increase in fuel for nuclear power. Dr. Gofman was the third chemist in the world to work with plutonium. His access to plutonium enabled Dr. Gofman to isolate plutonium, which was urgently needed by Dr. Oppenheimer on the Manhattan Project. Dr. Gofman has been especially noteworthy in recent years because of his knowledge of the hazards to human health as a result of exposure to radiation. He has written extensively in the scientific and medical treatises concerning the many characteristics and hazards of nuclear energy. His curriculum vitae was most impressive. In the *Silkwood* trial, Dr. Gofman testified as to the characteristics of plutonium, its uses and dangers to living organisms, and in particular to the consequences of the defendant's operation of the nuclear plant to the environment and human physiology and the damage done by the exposure of the plaintiff to radiation.

Dr. Gofman's testimony was followed by that of Dr. Morgan, whose background, experience and reputation was equally impressive. Dr. Morgan was involved in the Manhattan Project at the University of Chicago as one of five nuclear health physicists who were predominantly involved with the safety aspects in producing the world's first atomic bomb. Dr. Morgan worked at Oakridge National Laboratory with the Nobel Prize physicist, Dr. Ferme, and other famous scientists. He later became the director of health physics at Oakridge National Laboratories. Dr. Morgan's expertise and testimony was centered on the health physics aspects of the dangers and usage of nuclear energy. In fact, he was often referred to as the "Father of Health Physics" in nuclear field as both a scientist and educator. His curriculum vitae was superb.

Both of these eminent scholars and scientists were subjected to comprehensive and probing questions on direct examination by plaintiff's counsel, Gerald Spence, an excellent lawyer, and to probing and exhaustive examination by the very competent staff of attorneys representing the defendant Kerr-McGee. My recollection, buttressed by the written record of the case, of which I have a copy, indicates that their testimony was unsullied in its credibility aspects. In sum, the testimony was detailed, clear and convincing.

Experts are judged by the finders of fact—whether it be a jury or a judge—upon their background of experience, their learned treatises, their demeanor as a witness (in speaking authoritatively), and last and most important—on their ability to withstand exhausting and penetrating cross-examination. Cross-examination is the hallmark of our American justice system in the search for truth of the testimony of any witness—whether the case before the court be civil or criminal. A kindred quality in expert witnesses which I have noted through the years, and which impresses both the judge and jury, is their ability to be honest and open and to have the candor to discuss the pros and cons of their opinion—that is, acknowledging where error might lie, or other experts differ—and finally, to explain why their opinion is the sound one.

Reviewing in my own mind the many excellent experts in areas of human endeavor that I have heard during my twenty-four years on the federal trial bench, I believe these two gentlemen are among the most impressive, articulate, honest, and knowledgeable expert witnesses in their respective fields that I have had the privilege to hear and observe.

As a caveat to the purpose of this letter, I have been informed it may be used in connection with the preparation of a book by Dr. Morgan on the background, development of the nuclear age, the personalities involved, the public problems of the usage of nuclear energy, including potential dangers, and some of the leading cases involving litigation of issues concerning nuclear energy, of which the *Silkwood* case was one. Since that case is history, I am pleased to express my personal opinion of the two scientists as expert witnesses in that litigation.

Sincerely,
Frank G. Theis [signed]

NOTES

PREFACE

1. Radionuclide and other technical terms are defined in the glossary.

2. Many reactors, especially the pressurized water reactors (PWRs), may have a positive void coefficient caused by putting boron in the circulating water. This is done to enable the boron to act as a shim that keeps the reactor operating at a constant power level. It has a very high capture cross-section (σ_c = 3840 barn for B-10) compared, for example, with pure C-12 (σ_c = 0.0035 b). The B-10 enters into a rather unusual reaction with thermal neutrons:

$$^{10}_{5}B + ^{1}_{0}n \rightarrow ^{4}_{2}\alpha + ^{7}_{3}Li$$

where σ = 0.045 b for Li-7. In the reactor, the B-10 is "burned up" (converted to lithium, Li-7) as the reactor fuels U-235 and Pu-239 are consumed, so the reactor tends to operate at a fixed level without partial withdrawal of the control rods. This is an engineering advantage but a serious safety disadvantage, since losing reactor water is like jerking out the reactor control rods and causing an explosion. Both Three Mile Island and Chernobyl were PWRs.

PROLOGUE

1. K. Z. Morgan and Nielsen, "Shower Production"; Nielsen and K. Z. Morgan, "Absorption of Penetrating Component of Cosmic Radiation"; Nielsen, J. E. Morgan, and K. Z. Morgan, "Rossi Transition Curve"; Nielsen, Ryerson, Nordheim, and K. Z. Morgan, "Measurement of Mesotron Lifetime"; Nielsen, Ryerson, Nordheim, and K. Z. Morgan, "Differential Measurement of Meson Lifetime" (see bibliography for full citations).

2. Cosmic radiation consists of many types of radiation including the meson, which was discovered by cosmic ray researchers. The primary π-meson has a mass of 273.18 m_e where m_e (the mass of the electron) is 9.1094 x 10^{-28} gm. The proton has a mass of 1,836.15 m_e, and the neutron is slightly heavier, with a mass of 1,838.68 m_e. The primary cosmic radiation consists of protons and electrons. When these particles strike the earth's atmosphere, they produce neutrons and mesons consisting of muons and pions.

3. From another perspective one may consider that the universe is made up of three entities: photons, leptons, and hadrons. The present subdivisions are as follows:

PHOTONS Waves, which in order of decreasing energy are cosmic, gamma, X-ray, ultraviolet, visible light (violet, indigo, blue, green, yellow, orange, red), infrared, microwave, radio waves, TV waves, and so on.

LEPTONS e^- or β^- (electrons), e^+ or β^+ (positrons), ν_e, ν_μ, μ^+, μ^-, and so on.

HADRONS Mesons: π^-, π^0, π^+, η, K^0, K^+, K^- and 20 heavy resonate mesons with short halflife of 10^{-23} sec. Barons, nucleons (p^+, n), hyperons, Λ, Σ^+, Σ^0, Σ^-, Ξ^-, Ξ^0, and so on. A more recent theory of nuclear particles postulates that the hadrons are composed of six fundamental quarks.

In this book, electrons, protons, neutrons, and mesons are classed as basic particles. The superscripts indicate the electrical charge (^{o}, $^+$, $^-$).

CHAPTER ONE
The Genie Leaves the Bottle: The Manhattan Project

1. The first pile at the University of Chicago is now a historic site.

2. In some Manhattan Project sites, "Product" was referred to as "49." This number was the reverse of 94, the atomic number of plutonium. The code word "Product" was not used for Pu-239 after the Alamogordo test on July 16, 1945.

3. Actually, Meitner and her nephew Frisch were Austrian, but since Germany had absorbed Austria, they were technically Germans at that time.

4. The threshold skin erythema dose was an unsatisfactory unit. It was lower for young women than for husky men; it could be produced by ultraviolet radiation as well as by ionizing radiation; and the same amount of reddening was produced by less energy absorbed per gram of human soft tissue for low-energy radiation (e.g., 40 keV X or gamma radiation) than for high-energy radiation (e.g., 1 MeV X or gamma radiation). Other ionizing radiation dose units have been developed since 1943: the roentgen (R); the roentgen equivalent physical (rep); radiation, absorbed dose (rad); the rad equivalent man (rem); the gray (Gy); and the sievert (Sv). These units may be defined as follows:

1 roentgen = 2.58×10^{-4} coulombs per kilogram of air or 1 electrostatic unit (esu) per cubic centimeter of air at Normal Pressure Temperature (NPT) or 87.7 ergs/gram of air

1 rep = 2.58×10^{-4} coulombs per kilogram of anything or 87.7 ergs/gram of anything

1 rad = 100 ergs/gram of anything (0.01 joule per kilogram)

1 rem = 1 rad unit times modifying factors, such as specific ionization

1 gray = 100 rad

1 sievert = 100 rem

5. When a neutron collides with a proton, the proton, if initially at rest, goes off in the same direction as the neutron and has all the energy of the neutron. Since the neutron strikes the proton at a random angle, the proton on the average goes off with only half the energy of the neutron.

6. The roentgen equivalent physical, or rep, got its name because the roentgen was defined in terms of the secondary electron energy produced and absorbed in air; this new unit corresponded to the same amount of absorbed energy as that of the roentgen, but the energy could be produced and absorbed in anything. In this case, "anything" usually meant soft human tissue.

7. Bob Coveyou, also a health physicist at Chicago, was among those transferred to Tennessee, but shortly afterward he joined the Mathematics Division.

8. Maybe here is where Wollan made a wise decision and I the wrong one. I still ask myself, should I have left health physics and returned to cosmic radiation research? Wollan brought with him his X-ray spectrometer and modified it in such a way that he could diffract neutrons emitted from an experimental hole in the 6-feet-thick wall about our graphite reactor. In this work, he trained a young scientist in the diffraction technique. In 1985 that same scientist, C. Shull, received the Nobel Prize for his contributions to this field of physics and expressed regret that Ernie Wollan was no longer living and hence could not share in the honors.

9. U-233 is made in reactors from Th-232 by the reactions:

$$
\begin{aligned}
&{}^{232}_{90}\mathrm{Th} + {}^{1}_{0}n \rightarrow {}^{233}_{90}\mathrm{Th}\ (22.3m)\\
&{}^{232}_{90}\mathrm{Th} \rightarrow {}^{0}_{-1}\beta + {}^{233}_{91}\mathrm{Pa}\ (27d)\\
&{}^{233}_{91}\mathrm{Pa} \rightarrow {}^{0}_{-1}\beta + {}^{233}_{92}\mathrm{U}\ (1.59 \times 10^{5}y)
\end{aligned}
$$

It also is a fissile material used in some test weapons.

Pu-239 is made from U-238 in reactors by the reactions:

$$
\begin{aligned}
&{}^{238}_{92}\mathrm{U} + {}^{1}_{0}n \rightarrow {}^{239}_{92}\mathrm{U}\ (23.51m)\\
&{}^{239}_{92}\mathrm{U} \rightarrow {}^{0}_{-1}\beta + {}^{239}_{93}\mathrm{Np}\ (2.355d)\\
&{}^{239}_{93}\mathrm{Np} \rightarrow {}^{0}_{-1}\beta + {}^{239}_{94}\mathrm{Pu}\ (24{,}110y)
\end{aligned}
$$

The U-235 is a component of natural uranium (0.7196%).

10. The early piles (nuclear reactors) used a moderator, such as carbon, to slow down the neutrons and increase the probability that they would enter the nucleus of a U-235 atom and cause its fission with the production of 2 or 3 neutrons. Many of the neutrons, however, were captured or lost from fission. The probability for capture is the relative magnitude of the capture cross-section expressed in barns, b, and given as σ_c. Some values of σ_c are 0.0035 b for C-12, 2.6 b for Fe-56, 20.7 b for Co-59, and 0.0005 b for H-2. So C-12 or H-2 were good moderators, provided they were not contaminated with elements such as Co-59 or Fe-56.

11. One of the first things we did at Oak Ridge was to arrange for the manufacture of pure carbon—thanks to the guidance of scientists such as John Wheeler and Eugene Wigner.

12. The amount of hydrogen in some of the concrete shielding had been increased because the single proton that makes up the hydrogen nucleus is most efficient in slowing down fast neutrons by the billiard ball principle. (If two balls have exactly the same mass, and one strikes the other head on, then the first ball stops dead in its tracks and the second ball [except for friction losses] moves on with all the momentum and velocity of the striking ball.) The mass of a proton is

slightly less than the mass of a neutron. Thermal neutrons are readily captured by calcium (Ca-40 has a capture cross-section, $\sigma_c = 0.41$ b), a major constituent of concrete, and by impurities in the concrete, such as Na-23 ($\sigma_c = 0.43$ b), Cl-35 ($\sigma_c = 43.7$ b), Co-59 ($\sigma_c = 20.7$ b), B-10 ($\sigma_c = 3840$ b), and Cd-113 ($\sigma_c = 20{,}000$ b). For comparison, the capture cross-section for H-1 is 0.332 b; for H-2, 0.0005 b; and for C-12, 0.0035 b. Since the fast neutrons strike at random angles, on average only half the momentum is transferred at each collision. Thus, it is easy to see why heavy water (deuterium) containing H-2 is a prime choice as a neutron moderator (slowing-down substance) in a pile (or nuclear reactor) such as the Canadian CANDU reactor. In a pile (reactor), the fewer neutrons lost in the moderator, the less fuel is required for operations.

13. Quoted in Moore, "Incident at Stagg Field," 14–15.

14. When the plaques were in the pile, fast neutrons of energy greater than 5 MeV knocked protons out of the sulfur nuclei and resided in those nuclei in their place. Since the number of protons in the nucleus determines the element or the atomic number, this substitution changed sulfur to phosphorus by the reaction:

$$^{32}_{16}\mathrm{S} + {}^{1}_{0}n \rightarrow {}^{32}_{15}\mathrm{P} + {}^{1}_{1}\mathrm{p}$$

$$^{32}_{15}\mathrm{P}\,(14.28\mathrm{d}) \rightarrow {}^{0}_{-1}\beta + {}^{32}_{16}\mathrm{S}$$

15. Several other serious radiation accidents took place in the years after the early period, from 1951 until I left Oak Ridge in 1972, although no one suffered immediate detectable injury:

On October 4, 1957, an employee mistakenly entered the pipe tunnel room, where highly radioactive fission products were being flushed to the tank farm. He received a 63 rem dose.

On November 12, 1957, an employee entered a hot cell to mix chemicals and check the agitator without realizing some of the materials were fission products. He received a 13 rem dose.

During the period October 28–31, 1959, a leak of an evaporator in Building 3019 resulted in an estimated release of 2,000 curies to the holding ponds and ultimately the Clinch River.

On November 12, 1959, a leak of nearly 15 curies of "hot particles" resulted in an escape up the red brick stack.

On November 20, 1959, a chemical explosion in a steel tank in Building 3019 caused 40 microcuries of Pu-239 to be released into a nearby building.

16. High levels of radioactive gases, especially argon-41, were a constant problem with the reactor, one that was never satisfactorily resolved. Most of the Ar-41 was produced in neutron capture by argon gas (Ar-40), which makes up 0.943 percent of natural air:

$$^{40}_{18}\mathrm{Ar} + {}^{1}_{0}n \rightarrow {}^{41}_{18}\mathrm{Ar}\,(109.3m) \quad \left\{\begin{array}{l} 2.49\ MeV\ \beta \text{ at } 0.8\% \\ 1.98\ MeV\ \beta \\ 1.293\ \gamma \text{ at } 99\% \end{array}\right\}$$

CHAPTER TWO
The Genie's Anger Unleashed: The Truman Administration's Greatest Mistake

1. See Einstein to President Roosevelt, March 25, 1945, Harry S. Truman Library, Independence, Missouri; reprinted in appendix 4.

2. "The United States Strategic Bombing Survey: The Effects of Atomic Bombs on Hiroshima and Nagasaki," 22, Truman Library.

3. Donovan, *Conflict and Crisis: The Presidency of Harry S. Truman 1945-1948*, 71 (emphasis added).

4. Ibid.

5. Einstein to President Roosevelt, March 25, 1945, Truman Library (see appendix 4).

6. For reasons of security, Szilard was unable to share with Einstein the particulars of his recommendations.

7. *Harry S. Truman Encyclopedia*, G. K. Hall Presidential Encyclopedia Series, s.v. "Atomic and Hydrogen Bombs," 13; Truman Library.

8. Szilard to Matthew J. Connelly, August 17, 1945, Truman Library.

9. Notes of Interim Committee Meeting of May 31, 1945, Truman Library.

10. Notes of Interim Committee Meeting of June 1, 1945, Truman Library.

11. Science Panel: Recommendations on the Immediate Use of Nuclear Weapons, June 16, 1945, Truman Library.

12. Arthur H. Compton, memorandum to Colonel K. D. Nichols, July 24, 1945, Truman Library.

13. "The United States Strategic Bombing Survey: The Effects of Atomic Bombs on Hiroshima and Nagasaki," 22–23, Truman Library.

14. Farrington Daniels, memorandum to A. H. Compton, July 13, 1945, regarding "Poll on the Use of Weapon," Truman Library.

15. Although I signed several petitions of this nature, I do not know if my signature appears on this particular petition. The signatures in the Truman Library are incomplete for this particular document.

16. Arthur H. Compton, memorandum to Colonel K. D. Nichols, July 24, 1945, Truman Library.

17. Colonel K. D. Nichols, memorandum to Major General L. R. Groves, July 25, 1945, Truman Library.

18. Goldberg, "Groves Takes the Reins," 32–37.

19. AGWAR Message No. 41011, "To the President from the Secretary of War," July 30, 1945, with Truman, handwritten note to "Sec. War," Truman Library; reprinted in appendix 5.

20. Weinberg, *First Nuclear Era*, 269.

21. *Truman Encyclopedia*, s.v. "Atomic and Hydrogen Bombs," 13; Truman Library.

22. *Truman Encyclopedia*, s.v. "Atomic and Hydrogen Bombs," 14; Truman Library.

23. *Truman Encyclopedia*, s.v. "Atomic and Hydrogen Bombs," 15; Truman Library.

24. "The United States Strategic Bombing Survey: The Effects of Atomic Bombs on Hiroshima and Nagasaki," 23, Truman Library.

CHAPTER THREE
The Early Years at ORNL

1. Examples include measurement of gamma radiation from iodine-131 in the thyroid gland and bremsstrahlung produced by deceleration of the beta particles from strontium-90 in the skeleton.

2. Years later I discovered that our own Oak Ridge facilities constituted one source of this radioactive contamination. For example, a large 10-acre junkyard at Y-12 contained contaminated metals. Although I had no responsibility for Y-12, a considerable amount of the metals stored there had been contaminated at X-10. I managed to stretch my tight budget enough to permit a survey twice a week. The fenced junkyard held large stacks of stainless steel pipe, aluminum objects, and copper tubing, and at least an acre full of cast-iron objects. The stainless steel pipes contained such high levels of uranium and plutonium that they could never be decontaminated to a "safe" level. I was very disturbed when my health physics surveyors reported on several occasions that some of the piles of contaminated material had mysteriously disappeared. We complained to management but never learned what had happened to this contaminated metal.

3. Fortunately during my tenure at the laboratory exposures were low-level, and we experienced no deaths or permanently disabling injuries. Despite our ostensible success, however, a thorough analysis of death certificates and film badge readings fifty years later revealed a statistically significant increase in all types of cancer from even these exposures.

4. The alpha counter responded only to alpha radiation. It was necessary to know the alpha dose in a field of mixed types of radiation (X, γ, β, n, α) because alpha radiation measured in energy absorption units, rads or Gy, is 30 times more harmful than X, γ, or β radiation (1 rad of $\alpha = 30$ rem, or 1 Gy of $\alpha = 30$ Sv). The alpha counter was a specially designed proportional counter. Alpha radiation has a range of only a few inches in air, so measurements were made with the counter held close to the contaminated surface.

5. The Hurst neutron dosimeter was a specially designed proportional counter circuit that responded only to fast neutrons (energy above 1 MeV). Fission products decay by emission of α, β, γ, X rays, and electrons, but not by emission of neutrons. During a critical assembly of fissile material (Pu-239, U-235, or U-233), two or three neutrons are emitted per fission.

6. There had been dust and smoke in the area that carried fission products, uranium and plutonium. The face masks filtered out this dust.

7. The buildings and road near the site of the accident were contaminated with fission products and plutonium. The use of tar and paint to prevent resuspension was a very ingenious move by Clark.

8. The current inventories of plutonium in this facility did not indicate at any time enough for a critical assembly, but over a period of months many "missing amounts" and "traces" had added up to what could be a critical assembly.

9. As the Geiger counter was brought closer and closer to a source of ionizing radiation, clicks were produced faster and faster until the counter could not recover during counts (a low resolving time). Then the counts stopped and the pointer on its meter dropped to zero. Any competent health physics surveyor would realize this was an acutely dangerous situation.

10. While the U-235 in the barrel was in a critical assembly, radioactive fission products were produced. Apparently enough of the liquid had been discharged from the barrel some time later that there was no longer a critical assembly. At any moment, however, there was the risk that more U-235 solution might leak into the barrel and cause this miniature pile to become active again. Following a critical assembly of fissile material, most of the early radiation is from radionuclides with short half-lives. The activity decreases with time following a critical assembly lull.

11. The gamma activity of a fission mixture following the burst at critical assembly drops off roughly inversely with time, so if the gamma dose rate, for example, was 1,000 rem per hour at 1 hour after the criticality burst, it would be about 500 rem/hr at 2 hours, 250 rem/hr at 3 hours, and so forth.

12. At the time of the Y-12 accident, Hurst had not yet developed the method of analysis of hair for the activation of sulfur in the production of phosphorus-32 by neutrons of energy above 5 MeV. He developed this method one year later. Had this method been available, we could have made an accurate estimate of the fast neutron dose. This reaction is:

$$^{32}_{16}\mathrm{S} + ^{1}_{0}\mathrm{n} \rightarrow ^{1}_{1}\mathrm{p} + ^{32}_{15}\mathrm{P}\ (14.28\ \mathrm{d})$$

Instead, he measured the Na-24 produced in blood by thermal neutrons in the reaction:

$$^{23}_{11}\mathrm{Na} + ^{1}_{0}\mathrm{n} \rightarrow ^{24}_{11}\mathrm{Na}\ (14.96\ \mathrm{h})$$

Dose estimates for the eight high-exposed workers are presented in the table in the text.

13. These doses were determined from several sources, primarily from analysis of the blood samples, but also from experiments simulating the Y-12 accident. In our simulation, pigs served as stand-ins for the Y-12 humans who had not worn film badges.

14. But see note 3 above.

15. The Bragg-Gray Principle states that the dose of ionizing radiation relates to the electrical current in an air cavity in that material. This principle was used in designing standard air chambers used for accurate measurements of radiation dose. The air cavity had a central electrode, and a high voltage was applied between this

electrode and the walls of the cavity or chamber. The current produced was proportional to the dose of ionizing radiation to which this meter was exposed.

CHAPTER FOUR
My Biggest Mistake

1. The International Commission on Radiological Protection (ICRP) and the U.S. Atomic Energy Commission (AEC) had set as the maximum 0.02 μg of Pu-239 in the body for members of the public and 30 times this for radiation workers. I had shown that these values were still far too high and needed to be reduced at least by a factor of 200.

2. This was to be a unified system in which the fission products would be continuously removed from the circulating molten reactor salt and fuel (U-233) and incorporated in glass rods and steel casings ready for shipment to an abandoned salt mine as a permanent and safe resting place.

3. The U-233 fuel is produced in the blanket of the MSTB from natural thorium, Th-232, by the reactions:

$$^{232}_{90}\mathrm{Th} + ^{1}_{0}\mathrm{n} \rightarrow ^{233}_{90}\mathrm{Th}\ (22.3m)$$
$$^{233}_{90}\mathrm{Th} \rightarrow ^{0}_{-1}\beta + ^{233}_{91}\mathrm{Pa}\ (27.0d)$$
$$^{233}_{91}\mathrm{Pa} \rightarrow ^{0}_{-1}\beta + ^{233}_{92}\mathrm{U}\ (1.59 \times 10^{5}y)$$

The intensity of the gamma radiation or the specific activity (curies/gram) is given by the relation $\mathrm{Ci/gm} = 3.577 \times 10^{5}/(\mathrm{Ty\ W})$, where W = atomic weight (mass) and Ty = half-life in years.

In the U-233 cycle, a considerable amount of U-232 is produced in the reaction:

$$^{233}_{92}\mathrm{U} + ^{1}_{0}\mathrm{n} \rightarrow ^{2}_{0}2\mathrm{n} + ^{232}_{92}\mathrm{U}$$

This is similar to the Pu production of Pu-238 in the Pu-239 cycle:

$$^{239}_{94}\mathrm{Pu} + ^{1}_{0}\mathrm{n} \rightarrow ^{2}_{0}2\mathrm{n} + ^{238}_{94}\mathrm{Pu}$$

The U-232 (68.9 y) has a relatively short half-life of 68.9 years, and it and its daughter products are extremely radioactive because of this short half-life as follows (* = 64% of the time, ** = 36% of the time):

$$^{232}_{92}\mathrm{U}\ (68.9y) \rightarrow ^{4}_{2}\alpha + \gamma + ^{228}_{90}\mathrm{Th}\ (1.913y)$$
$$^{228}_{90}\mathrm{Th} \rightarrow ^{4}_{2}\alpha + \gamma + ^{224}_{88}\mathrm{Ra}\ (3.66d)$$
$$^{224}_{88}\mathrm{Ra} \rightarrow ^{4}_{2}\alpha + \gamma + ^{220}_{86}\mathrm{Rn}\ (55.6s)$$
$$^{220}_{86}\mathrm{Rn} \rightarrow ^{4}_{2}\alpha + \gamma + ^{216}_{84}\mathrm{Po}\ (0.145s)$$
$$^{216}_{84}\mathrm{Po} \rightarrow ^{4}_{2}\alpha + \gamma + ^{212}_{82}\mathrm{Pb}\ (10.64h)$$
$$^{212}_{82}\mathrm{Pb} \rightarrow ^{0}_{-1}\beta + \gamma + ^{212}_{83}\mathrm{Bi}\ (1.009h)$$
$$^{212}_{83}\mathrm{Bi} \nearrow ^{4}_{2}\alpha + \gamma + ^{208}_{81}\mathrm{Tl}\ (3.053m)\ *$$
$$^{212}_{83}\mathrm{Bi} \searrow ^{0}_{-1}\beta + \gamma + ^{212}_{84}\mathrm{Po}\ (2.98 \times 10^{-7}s)\ **$$

$$^{208}_{81}\mathrm{Tl} \rightarrow {}^{0}_{-1}\beta + \gamma + {}^{208}_{82}\mathrm{Pb}\ (\textit{Stable})$$

$$^{212}_{84}\mathrm{Po} \rightarrow {}^{4}_{2}\alpha + \gamma + {}^{208}_{82}\mathrm{Pb}\ (\textit{Stable})$$

In the case of the Pu-239 weapon production, the radiation background is much less than for the U-233 weapon because of the long half-lives:

$$^{239}_{94}\mathrm{Pu}\ (24{,}110y),\ {}^{240}_{94}\mathrm{Pu}\ (6{,}537y),\ \textit{and the chain}$$

$$^{238}_{94}\mathrm{Pu}\ (87.74y) \rightarrow {}^{4}_{2}\alpha + \gamma + {}^{234}_{92}\mathrm{U}\ (2.45 \times 10^{5}y)$$

The X-ray and gamma background radiation of U-233 and its carriers is very high compared with that of Pu-239 or U-235 and their carriers, primarily because of the short half-life of U-233 and its carriers and daughter products. Specific activity, Ci/g, is inversely proportional to half-life, T, as given by:

$$\mathrm{Ci/gm} = 3.577 \times 10^{5}/(\mathrm{Ty\ W})$$

where W = atomic weight and Ty = years. For example, we have in the U-233 cycle half-lives:

U-233 (1.59×10^5 y) → Th-229 (7,900 y) → Ra-225 (14.9 d) → Ac-225 (10 d) → Fr-221 (4.8 m) → At-217 (0.032 s) → Bi-213 (45.6 m) → Po-213 (4μs) → Pb-209 (3.25 h) → Bi-209 (stable)

and in the Pu-239 cycle:

Pu-239 (24,110 y) → U-235 (7.04×10^8 y) → Th-231 (1.063 d) → Pa-231 (3.25×10^4 y) → Ac-227 (21.77 y) → Th-227 (18.72 d) → Ra-223 (11.43 d) → Rn-219 (3.96 s) → Po-215 (1.78 ms) → Pb-211 (36.1 m) → Bi-211 (2.14 m) → Po-211 (0.516 s) → Pb-207 (stable)

The 7.04×10^8 y in the Pu-239 cycle effectively removes subsequent daughters that otherwise would accumulate in Pu-239 fuel, making Pu-239 fuel less radioactive than U-233 fuel.

4. Spontaneous fission is a problem in constructing a nuclear weapon because neutrons are produced in fission and tend to predetonate the fissile material and blow it apart unfissioned before it is compacted in a tight, high density, supercritical assembly. Thus, the shorter the spontaneous fission half-life, the greater the predetonation problem.

5. Predetonation can be caused by a cloud of neutrons as the assembly approaches a critical arrangement.

6. The PWR and BWR became the power reactors of choice in the United States. Unfortunately the nuclear industry took this step before at least fifty years of research to determine whether reactors could be made "safe" and economical. We compounded our error in judgment by failing to first determine satisfactory methods of waste disposal and reactor decommissioning.

7. When I returned to ORNL, my friends at the laboratory told me how over a few days the course had been changed, and money and jobs had become more important than a safer and cheaper reactor system. They were happy not to lose their jobs, of course, but scornful of a weak management. They did not appreciate or understand why Culler, rather than Weinberg, seemed to be running the lab.

8. The following year I gave portions of what was originally in my Neuherberg paper at an obscure health physics training program in Yugoslavia. My contribution was published in the program proceedings, but I fear that no one beyond this little group in Yugoslavia ever read those proceedings or heard my warnings. Even Weinberg must have felt pressure, since afterward he called me to his office to say, "Karl, your Yugoslavian paper did not reach the high standards of your many other publications."

9. The U.S. government agency responsible for atomic energy at ORNL has gone through a number of changes. First it was the Army Corps of Engineers, then the AEC, which lasted for thirty years. Next came the Energy Research and Development Administration (ERDA) and the Nuclear Regulatory Commission (NRC). Later, most of ERDA's functions were taken over by the DOE, with a few given to the NRC.

10. Weinberg, *The First Nuclear Era*, 199–200.

11. Had I been so honored I would have declined. I was anxious to get back to teaching and conducting research.

12. France is the only country that has accepted the fast breeder dream and built both the Phoenix and Super Phoenix fast breeder reactors that Weinberg and I opposed, each in our own way. They have been a continuous national headache, an endless money drain, and, to many people like me, a simmering caldron gathering steam for the day it might beat the Chernobyl world record.

13. Morgan, "Ionizing Radiation Exposure."

14. M. Sohrabi and Gary Stillwagon proved to be two of my most promising Ph.D. students. Stillwagon later added an M.D. degree to the Ph.D. he received at Georgia Tech.

15. Morgan, "Body Burden of Long-Lived Isotopes"; Morgan, Snyder, and Ford, "Relative Hazard of Various Radioactive Materials"; Morgan, "Suggested Reduction of Permissible Exposure to Plutonium and Other Transuranium Elements."

16. Klaus Becker, a visiting scientist from Germany, and M. Sohrabi, a visiting scientist from Iran and my former student, were instrumental in this early research.

17. Stillwagon and Morgan, "In-Situ Dosimetry of Plutonium-239."

18. The universally used photon track dosimeter has several flaws for measurement of fast neutron dose: (1) it requires a skilled technician to count the proton tracks using a dark field microscope; (2) it is very insensitive and considerable skill is required to measure a fast neutron exposure that is one-half the permissible amount; and (3) under conditions of high temperature and humidity, half the proton tracks can fade beyond detection in two weeks' time. It was I who, at the suggestion of Lyle Borst, developed and introduced this proton track dosimeter. Perhaps I can be forgiven because it was the best known method at that time and—until K. Becher, Sohrabi, Stillwagon, and others introduced the polycarbonate electrochemical etching method two decades later—it was all that was available for personnel monitoring of fast neutron dose.

19. I would be remiss if I neglected to recall here an important publication by Larsen and Oldham, "Plutonium in Drinking Water." They administered Pu-239 to rats in drinking water used by the public in the city of Chicago and explored the effect of oxidation state of plutonium on body uptake from the gastrointestinal tract. They made what to me, as a physicist and not a chemist, was the startling discovery that the percent uptake of Pu-239 to the skeleton and liver when the Pu-239 was in the VI oxidation state was 1.75 percent, compared with an uptake of only 0.001 percent when in the IV state.

The MPC values of the ICRP, NCRP, DOE, and NRC have all been based on the erroneous assumption that the uptake is only 0.003 percent (three times the IV state).

In their studies, Larsen and Oldham varied the amount of chlorination in the Chicago drinking water. At very low chlorination most of the plutonium was in the IV oxidation state, but as the chlorination was increased to 1 part per million and above (most of Chicago drinking water was at 1 part per million or above), the oxidation increased rapidly, changing the IV state to the VI state. The presence of food in the gastrointestinal tract changed it from the VI to the IV state. The ratio of the two percentages is 1.75/0.003 = 583. This means the MPC values given in ICRP-2 (1959) and NCRP-69 (1959) were too high for plutonium by a factor of 583. In ICRP-30 (1988), the ICRP reduced this discrepancy from a factor of 583 to a factor of 12, but in ICRP-61 (1990), the discrepancy was raised to a factor of 23, where it stands at present.

Plutonium is by no means the only radionuclide that has been juggled this way by the ICRP. For example, the ICRP factors of increase in the MPC level for Sr-90 in water above the ICRP-2 (1959) value (when I was chairman of its internal dose committee) have been ICRP-30/ICRP-2 = 25 and ICRP-61/ICRP-2 = 1.5. I was so concerned about these increases and the juggling of MPC values by both the ICRP and the NCRP that I published several papers in which I deplored this method of operation of standard-setting bodies. See Morgan, "Health Physics: Its Development, Successes, Failures and Eccentricities"; Morgan, "Do Low-Level Radiation Health Data Justify Fear or Contribute to Phobia?"; Morgan, "A Juggling Act"; Morgan, "Changes in International Radiation Protection Standards"; and Morgan, "The International Commission on Radiological Protection Made a Bad Mistake."

20. Following the Three Mile Island reactor accident, a hearing was held before Judge Rambo in Harrisburg, Pennsylvania. David Berger was the lawyer for the plaintiffs. V. P. Bond was the expert witness for the defense, and I the expert witness for the plaintiffs. In the settlement we won what has amounted to about $12 million (including interest). These funds have been used to support independent research on dosimetry, instrumentation, and cancer incidence among former AEC contract workers at Oak Ridge, Hanford, Los Alamos, and other sites, in relation to the film badge readings of radiation exposure. After the hearing, Berger set up a committee to administer contracts for these research programs, and I was

asked to serve as chair. Some very useful reports have been published under the sponsorship of this committee. A few copies of these reports are still available by writing to TMIPHFC at 1622 Locust St., Philadelphia, PA 19103.

21. See September 25, 1962, declassified letter of the deputy director of the EPA predecessor, in chapter 7. The document shows how to "stack the deck" when a government agency solicits the opinion of a scientist.

CHAPTER FIVE
Nuclear Incidents in Other Facilities

1. Bismuth-209 was placed in a reactor where it captured a neutron and became Bi-210, which has a short half-life of 5.01 days. During the decay of Bi-210, a beta particle is emitted, producing polonium-210, which has a half-life of 138.4 days. These reactions can be given as:

$$^{209}_{83}\text{Bi}\ (\textit{Stable}) + ^{1}_{0}\text{n} \rightarrow ^{210}_{83}\text{Bi}\ (5.01d)$$

$$^{210}_{83}\text{Bi} \rightarrow ^{0}_{-1}\beta + ^{210}_{84}\text{Po}\ (138.4d)$$

$$^{210}_{84}\text{Po} \rightarrow ^{206}_{82}\text{Pb}\ (\textit{Stable}) + ^{4}_{2}\alpha$$

$$\downarrow$$

$$^{4}_{2}\alpha + ^{9}_{4}\text{Be}\ (\textit{Stable}) \rightarrow ^{12}_{6}\text{C} + ^{1}_{0}\text{n}$$

Devices containing Po-210 were the principal component of "triggers" used in our nuclear weapons: When alpha radiation from Po-210 strikes a light element, such as beryllium, it knocks out neutrons, which "trigger" the imploding U-235 or Pu-239. Therefore, any mention of Po-210 or Bi-209 or their use was prohibited and considered a violation of secrecy restrictions.

The fact usually is not appreciated that an explosion of a weapon is not just the release of a large amount of energy, but the release of this energy in a short time. Otherwise, the exploding weapon might "fizzle." Ideally one would like to make this time approach zero, because then energy/time approaches infinity. A Pu-239 weapon was far more difficult to build than a U-235 weapon, because the process of making Pu-239 in a reactor also produced a considerable amount of both Pu-240 and Pu-238, whose removal was extremely difficult and costly—much more so than separating U-235 from the natural mixture of U-234, U-235, and U-238. Pu-240 and Pu-238 were serious contaminants because they naturally emitted neutrons that would tend to predetonate a weapon if one were to use two slightly subcritical pieces of Pu-239 simply brought together in cannonlike detonation, similar to the process used in the U-235 weapon dropped over Hiroshima. The scientists at Los Alamos solved this problem by making the plutonium weapon an implosion type, where the supercritical mass of plutonium was compacted together in an extremely short time (billionths of a second). Also, during the implosion Po-210 was mixed with a light element, such as beryllium, so that the alpha radiation of Po-210 striking the light element produced instantaneously an intense cloud of neutrons in the compacted plutonium.

2. Wet filters are very inefficient because wetness blocks air flow through them. It is doubtful whether the large bismuth particles actually did much to improve the filtering of fission products.

3. I suppose that by now the Grim Reaper has in his way absolved the guilty persons in the top echelons of the British establishment who knew about the extreme danger of this alpha exposure (and had top-security clearance) but failed to inform Greg Marley so he could provide protection to thousands of persons downwind of Windscale. Also, the short half-life of Po-210 quickly erased the physical evidence.

4. The residual Po-210 after thirty-seven years is approximately 4×10^{-28} percent of its original activity.

5. Dose values were determined by Hurst. See Morgan and Turner, eds., *Principles of Radiation Protection*, 45.

6. The characteristic of Geiger counters to read zero in very high and dangerous radiation areas must be kept in mind by everyone using this instrument.

7. Thermal neutrons convert normal sodium (Na-23) into radioactive sodium (Na-24) by the reaction $^{23}Na + {}^{1}n \rightarrow {}^{24}Na$ (14.96 h).

8. Boron, B-10, is one of the tramp poisons that gobble up neutrons, $^{10}B + {}^{1}n \rightarrow {}^{11}B$. The thermal neutron capture cross-section of B-10 is $\sigma_c = 3740$ b, while that of B-11 is only $\sigma_c = 0.005$ b.

9. Cadmium has a large capture cross-section for thermal neutrons: $\sigma_c =$ 20,600 b for Cd-113, $\sigma_c = 0.04$ b for Cd-114. Converting Cd-113 to Cd-114 greatly reduces the neutron loss in the reactor.

10. Hawkes et al., *Chernobyl: The End of the Nuclear Dream*, 99.

11. Ibid.

12. Ibid., 101–2.

13. Ibid., 102.

14. Ibid., 133.

15. Medvedev, *Truth about Chernobyl*, 205.

16. Ibid., 201–2.

17. Hawkes et al., *Chernobyl: The End of the Nuclear Dream*, 195.

CHAPTER SIX
The Price

1. Joseph Rotblat, who was awarded the Nobel Prize for peace in 1995, reported in 1985 that his reanalysis of the data on the Japanese survivors of the two atomic bombs caused him to conclude that the mid-lethal dose for human beings was 154 rad. See International Physicians for Prevention of Nuclear War, 1985 Nobel Peace Prize Oslo Report, 27. The mid-lethal dose is the dose expected to kill half of the exposed persons in less than a week following exposure.

2. Weinberg, *First Nuclear Era*, 63.

3. We made some progress on how to handle radioactive waste during my twenty-nine years at Oak Ridge. When I left in 1972, responsibility for radioactive waste was transferred to ORNL Engineering Division. Since then the division has essentially been busy reinventing the wheel.

4. In 1943 there were no data on the lethal dose of ionizing radiation exposure of human beings and very little such data on animals.

5. Today the ICRP assumes that on the basis of energy absorption by the human body (as expressed in rad or gray units), the neutron is 30 times more harmful than gamma, X, or beta radiation. Thus a dose of 10 rads (or 0.1 Gy) of fast neutrons corresponds to a 300 rem (or 3 Sv) dose or is considered to be equivalent to 300 rads or 3 Gy gamma dose.

6. It was assumed that no one would drink water from the muddy White Oak Lake, but a person might ignore the "no fishing" signs and receive as much as 100 R external exposure in a day.

7. IXRPC was a forerunner of the ICRP. It was first organized in 1928 and was composed of seven members. It operated until 1937 but ceased to function during World War II. I was one of its members when it reorganized with thirteen members in 1950 as the ICRP. The U.S. National Council on Radiation Protection (NCRP) recommended a limit of 0.1 R/day or 0.5 R/week in 1934, and this was continued until 1949 when it recommended 0.3 R/week (actually 0.3 rem/week). In 1956 ICRP, and in 1957 NCRP, recommended 5 rem/year. In 1990 the ICRP reduced the occupational level from 5 rem/year to 2 rem/year averaged over any five-year period, and its level for members of the public was reduced from 0.5 rem/year to 0.1 rem/year.

8. Concentration factor = μCi/g fish tissue $\div$ μCi/g river water.

9. For example, in twenty years more than 99.999 percent of iodine-131 (8.04 d) would have been converted into stable xenon gas, Xe-131; 37 percent of cesium-137 (30.2 y) into stable barium, Ba-137; and 0.057 percent of Pu-239 (24,110 y) into U-235 (7.04×10^8 y) by the following transitions:

$$^{131}_{53}\text{I} \rightarrow {}^{0}_{-1}\beta + {}^{131}_{54}\text{Xe}\ (\textit{stable})$$
$$^{137}_{55}\text{Cs} \rightarrow {}^{0}_{-1}\beta + {}^{137}_{56}\text{Ba}\ (\textit{stable})$$
$$^{239}_{94}\text{Pu} \rightarrow {}^{4}_{2}\alpha + {}^{235}_{92}\text{U}\ (7.04 \times 10^8 y)$$

These calculations were made by use of the equation $I/I_0 = e^{-\lambda t} = e^{-0.693t/T}$, where t = 20 years and T is the half-life of the radionuclide.

10. When natural sulfur (S-32) is struck by a fast neutron (n), the neutron knocks a proton (p) out of the S-32 nucleus and takes its place, i.e. $^{32}S + {}^{1}n \rightarrow {}^{32}P + {}^{1}p$.

11. James Thomas, "A Summary of Radiation Releases from Hanford" (May 18, 1989), 4 (quoting March 1961 report of Columbia River Advisory Group, a committee formed by the AEC), 1.

12. Griffiths and Ballantine, *Silent Slaughter*, 114–15.

13. Ibid., 116.

14. Ibid., 117–18.

15. Ibid.

16. K. Z. Morgan, letter to *Oak Ridger*, December 19, 1993; Morgan, "Human Radiation Studies," 13.

17. Litton, "What Has America Done."

18. Ibid.

19. U.S. Senate, Committee on Governmental Affairs, "Human Radiation and Other Scientific Experiments," 101.

20. Ibid., 126, 129.

21. Ibid., 146–47.

22. "Interim Report of the Advisory Committee on Human Radiation Experiments," vi.

23. Ibid., 38.

24. Gallagher, *American Ground Zero*, xv.

25. Lauran Neergaard, "Cancer May Be Legacy of Fallout," *Wichita Eagle*, August 2, 1997.

26. "Interim Report of the Advisory Committee on Human Radiation Experiments," 17.

27. U.S. Senate, Committee on Governmental Affairs, "Human Radiation and Other Scientific Experiments," 100.

28. "Interim Report of the Advisory Committee on Human Radiation Experiments," xvi.

CHAPTER SEVEN
The Advance and Decline of Health Physics

1. Some copies of these instructional materials have been collected and are being preserved in the University Libraries Special Collections, Hoskins Library of the University of Tennessee, Knoxville. For a listing of some of the publications mentioned here, see bibliography.

2. In addition to the women scientist, other women workers made essential contributions to the development of health physics in other capacities. I would be remiss not to mention the contributions of Peggy Oldham, Marie Wright, and Jeanne Carver, who steadfastly served me in a secretarial capacity for years. Further, scores of "meter girls"—as they were called back then—painstakingly read thousands of personnel monitoring meters each evening. Natalie Tarr deserves recognition for her role as second mother to our students and later as my secretary for our new journal, *Health Physics*.

3. The handbooks of greatest international importance of Committee II were "Report of Committee II on Permissible Dose for Internal Radiation" (1959) and "Recommendations of the International Commission on Radiological Protection" (1962). The handbooks of greatest national importance were "Maximum Permissible Amounts of Radioisotopes in the Human Body and Maximum Permissible Concentrations in Air and Water" (1953) and "Maximum Permissible Body

Burdens and Maximum Permissible Concentrations of Radionuclides in Air and in Water for Occupational Exposure" (1959).

4. At the first general assembly in Rome, on September 7, 1966, the following societies were recognized as affiliates:

Argentine Radiation Protection Association

Association Belge de Radioprotection

Associazione Italiana di Fisica Sanitaria e di Protezione Contro le Radiazioni

Association Luxembourgeoise de Radioprotection

Asociacion Mexicana de Proteccion Radiologica y Fisica Medica

British Radiological Protection Association

Eotvos Lorand Physical Society, Health Physics Section, Hungary

Fachverband für Strahlenschutz (West Germany, Switzerland, and Austria)

Health Physics Society (United States and Canada)

Israel Health Physics Society

Japan Health Physics Society

Nederlandse Vereniging voor Stralingshygiene

Nordiska Sallskapet for Stralskydd (Norway, Sweden, Denmark, Finland, and Iceland)

Philippine Radiation Protection Association

Société Française de Radioprotection

At the Second International Congress on Radiation Protection, held in Brighton, England, May 3–8, 1970, these additional societies were recognized as affiliates:

Ōsterreichischen Verband fur Strahlenschutz (Austria)

Yugoslav Society for Radiological Protection

Czechoslovak Society of Nuclear Medicine and Radiation Hygiene

Indian Association for Radiation Protection

South African Association of Physicists in Medicine and Biology

Health Physics Section of the Polish Medical Physical Society

5. Until the publications of Muller and Stewart, the medical profession in general had assumed the levels of maximum permissible exposure to ionizing radiations that had been set by ICRP and NCRP were completely safe. The so-called tolerance limit in use, and the level I first set at Clinton Laboratories in 1943, was 0.1 r per day (36,500 mr/y). Stewart's research showed that doses of 500–600 mr increased the risk of the child developing cancer before the age of ten, and Muller's studies on flies indicated that there probably was no safe level of exposure relative to genetic damage. The present maximum permissible exposure level set by ICRP for members of the public is 100 mrem/y. Thus, the level used before 1943 was

365 times the present limit. See Stewart, Webb, Giles, and Hewitt, "Malignant Disease in Childhood and Diagnostic Irradiation in Utero."

6. *Silkwood v. Kerr-McGee Corp.,* 485 F. Supp. 566 (W.D. Okla. 1979).

7. Many hard-working and justly proud health physicists have not reached the "firing line" and been put to the test of testifying in court on behalf of an employee injured by workplace radiation exposure. Except for a handful, those health physicists who have appeared in court have essentially disregarded our professional obligations. Without public awareness and pressure, I fear that health physics will merely become an instrument of the nuclear-industrial complex.

8. The problem was significantly improved when the source of the contaminated iron oxide, iron in the ducts, was replaced by stainless steel.

9. Tamplin and Cochran, "A Report on the Inadequacy of Existing Radiation Protection Standards" and "The Hot Particle Issue."

10. H. Lisco, M. P. Finkel, and A. M. Brues, "Carcinogenic Properties of Radioactive Fission Products and of Plutonium," paper prepared for publication in the Plutonium Project Record of the Manhattan Project and presented at the 32nd Annual Meeting of the Radiological Society of North America, Chicago, 1946.

11. This statement was published in the Membership Handbook of the Health Physics Society, 1971–72 edition, xi.

12. Much of the support for the nuclear industry by members of ICRP related to the fact that they were employed or partly financed by the U.S. DOE. For example, some ICRP members obtained funds from the DOE for support of their research or that of their students, and they were hesitant to go on record as opposing programs of the agency that funded their research.

13. The quality factor is used to convert rads to rems or grays to sieverts. In the case of alpha radiation, this factor is 30 and is often taken as 30 for fast neutrons and 5 for thermal neutrons.

14. See my publications: Morgan, "The International Commission on Radiological Protection Made a Bad Mistake"; "Health Physics: Its Development, Successes, Failures, and Eccentricities"; "Changes in International Radiation Protection Standards"; "Do Low-Level Radiation Health Data Justify Fear or Contribute to Phobia?"; "A Juggling Act"; and "ICRP Risk Estimates—An Alternative View."

15. One person rem corresponds to one rem delivered to one person, 0.01 rem to 100 persons, 0.001 mrem to 1,000 persons, and so forth.

16. Morgan, Snyder and Ford, "Relative Hazard of Various Radioactive Materials."

17. See the ICRP report "Statement by the International Commission on Radiological Protection," ICRP/87/C:MC25, 2nd draft, Como, Italy, September 7–17, 1987.

CHAPTER EIGHT
Ecology and Nuclear Waste Disposal Studies

1. Although not technically "classified," these reports are nonetheless kept behind lock and key at ORNL. This is difficult to justify, since hundreds of copies were distributed to scientists around the world at the time of publication.

2. *Health Physics Research at Oak Ridge National Laboratory*, 14.

CHAPTER NINE
The Genie Goes to Court

1. *Silkwood v. Kerr-McGee Corporation*, 485 F. Supp. 566 (W.D. Okla. 1979); *Allen v. United States*, 588 F. Supp. 247 (D. Utah 1984).

2. *Silkwood* trial proceedings.

3. Spence and Polk, *Gerry Spence Gunning for Justice*, 130.

4. My coauthor advises me that Gerry Spence is one of the best trial lawyers in the world.

5. Spence and Polk, *Gerry Spence Gunning for Justice*, 153.

6. Page numbers cited parenthetically in this section of the text refer to the transcript of the *Silkwood* trial proceedings.

7. The full text of Judge Theis's appraisal is reproduced in appendix 6.

8. Because of the enormous size of the *Allen* case, the claims of twenty-four plaintiffs were selected as "bellwether" cases, typical of those that would follow. In ten of the twenty-four, the judge found that the plaintiffs' injuries were caused by excessive radiation fallout.

9. Under the Federal Torts Claims Act, private citizens are precluded from recovering punitive damages against the United States.

10. Page numbers cited parenthetically in this section of the text refer to the transcript of the *Allen* trial proceedings.

11. Knapp eventually published his work.

12. *Allen v. United States*, 588 F. Supp. 247, 419 (D. Utah 1984) (emphasis in original).

13. Letter to Ken Peterson, January 10, 1992.

14. *Allen v. United States*, 588 F. Supp. 247, 447 (D. Utah 1984).

15. *Allen v. United States*, 816 F.2d 1417 (10th Cir. 1987).

16. Ibid., 1427.

CHAPTER TEN
A Time for Reflection and Resolution

1. The term originates with George W. Ping and Penelope Canan, *SLAPPs: Getting Sued for Speaking Out* (Philadelphia: Temple University Press, 1996).

2. Lowe, "The Price of Speaking Out," 48.

3. Ibid., 48–50.

4. Ibid., 50.

5. Ibid., 52. California, Delaware, Massachusetts, Minnesota, Nebraska, Nevada, New York, Rhode Island, and Washington state all have such laws.

6. Report of the Monitored Retrievable Storage Review Commission, "Nuclear Waste: Is There a Need for Federal Interim Storage?" (1989).

7. Cochran and Norris, "A First Look at the Soviet Bomb Complex," 29.

8. Von Hipple was a member of the Three Mile Island Public Health Fund Committee.

9. National Public Radio, "Morning Edition," April 16–19, 1996. Further information and quotations in this section are drawn from this same source.

GLOSSARY

Acute radiation damage. Damage that is recognized shortly or within a few weeks following radiation exposure.

Air-cooled, graphite-moderated reactor (AGR). A nuclear reactor in which the fissile material, such as uranium or plutonium, is interspaced with graphite so that the carbon of the graphite can slow down or moderate the velocity of the neutrons that are formed during fission.

Alpha particle (α). One of the fundamental particles of matter. It is positively charged, consisting of two positively charged protons and two neutrons. It is identical with the nucleus of a helium atom and is emitted by many of the radioactive substances.

Alpha particle emitters. Radionuclides that emit alpha particles during radioactive decay.

Atomic (mass). The number of protons and neutrons in the atomic nucleus of an element.

Barn (b). A unit of nuclear cross-section. One barn equals 10^{-28} m^2 or 10^{-24} cm^2 = $1/10^{24}$ cm^2. The unit got its name from the image of a person trying to hit a barn door with a ball; in this case the "ball" is a nuclear particle, such as a neutron. The larger the barn door, the easier it is to hit. For many common substances the cross-section, σ, is not far from unity. For example, for thermal neutrons σ = 1.83 b for N-14, 0.43 b for Na-23, and 0.332 b for H-1.

Beta particle (β). An electron emitted from the nucleus of an atom. Since there are no beta particles in the nucleus of an atom, the particle is created at the moment of emission.

Borated water. Water containing borax or boric acid, used to capture neutrons or to reduce the neutron flux in a nuclear reactor. It is also used in some reactor control rods.

Boron (B). Chemical element of atomic number 5. It is of interest and often of concern to the nuclear physicist because of the large capture cross-section, σ_c, of one of its isotopes, B-10, where σ_c = 3840 b for thermal neutrons. B-10 has a natural abundance in boron of 19.9 percent.

Bragg-Gray effect. An effect by which radiation dose in a medium can be measured accurately, named for British scientist L. Hal Gray. Gray found that radiation dose to a tissue can be determined by measuring the ionization in a small air cavity in an instrument whose walls are made of tissuelike material. *See also*: Gray unit.

Bremsstrahlung. Electromagnetic radiation given off by deceleration of an electrically charged particle, such as a beta particle or a secondary electron produced by gamma radiation. When an intense radiation source, such as an operating nuclear reactor or 1000 Ci of Co-60, is placed under water, bremsstrahlung can be seen as a blue glow near the source. Also known as the Cerenkov effect.

Cadmium (Cd). The element with atomic number 48. The Cd isotope of atomic mass 113 (Cd-113) has a large capture cross-section, $\sigma_c = 20{,}600$ b, so it can be incorporated in steel rods used as reactor control rods.

Capture cross-section. *See*: Cross-section.

Chronic low-level radiation exposure. Exposure over a long period of time at a low dose rate, for example, 500 mrem per year (approximately five times the average natural background radiation level in the United States).

Concentration factor. The amount (concentration) of radionuclides that accumulate in tissue.

Control rod. A rod or bar of metal containing an element, such as boron or cadmium, with a large capture cross-section, σ_c, for neutrons. It is used to control the power level of a nuclear reactor or to shut off a reactor.

Cooling tower. A structure in which the water of a nuclear reactor is cooled as the heat is transferred to steam and air. In operation it usually emits a cloud of steam.

Cosmic radiation. A constant background of radiation from our own and distant galaxies that impinges on the earth equally from all directions of space. The primary cosmic radiation varies less than 1 percent in time and has the same intensity in all directions. This primary cosmic radiation strikes the earth's atmosphere with a frequency of about 1 particle per square centimeter per second and is composed mostly of protons. The energies of cosmic primary radiation range between 10^8 and 10^{13} MeV. Cosmic rays were discovered by Hess in 1911. Secondary cosmic radiation, produced in collisions of the primary cosmic radiation, is a mixture of mesons, electrons, and gamma radiation.

Coulomb. Unit for measuring the quantity of an electric charge, named for French physicist Charles de Coulomb (1736–1806). All matter is electrical in nature and consists of particles called protons and electrons. A proton has a positive charge, while an electron has a negative charge. It is possible to induce net positive or negative charges on matter. It is an elementary law of physics that charged bodies exert forces on each other. Bodies with similar charges repel one another, while those with unlike charges attract each other. Coulomb described the relationship between the magnitude of these electrical forces.

Critical assembly. An assembly of sufficient fissile material, such as U-235 or Pu-239, that is capable of sustaining a fission chain reaction. A chain reaction just maintains itself if the number of daughter neutrons produced is balanced by the number of neutrons that escape. A sphere is the optimum (smallest) arrangement of a critical assembly. The size of a critical assembly can be reduced by proper positioning of neutron-moderating and -reflecting materials, such as graphite or water, or by a nearby neutron source such as a radium-beryllium mixture. If the number of neutrons at a given instant in the assembly exceeds the number that escape, the assembly becomes a supercritical assembly, produces a blue flash of light, and breaks apart with a bang and the emission of an intense blast of neutrons and gamma radiation.

Criticality. That point in time when a nuclear chain reaction becomes self-sustaining.

Cross-section. The effective area presented by the nucleus of an element when struck by a fast-moving nuclear particle, such as a neutron, proton, or electron. The nucleus of the struck atom is symbolized as a target with a certain cross-sectional area, and the probability or likelihood of its being struck is directly proportional to this area. A number of things may happen in a nuclear collision: the struck atom may simply recoil; it might capture the striking neutron; or it might cause spallation of the struck nucleus, a reaction in which a large number of nucleons (particles) is released. In a very high-energy encounter a shower of nuclear particles may result. In this book we discuss only the case of striking neutrons that are captured by the target nucleus of the target atom—the capture cross-section, or σ_c.

Curie (Ci). Unit of measurement representing the activity required to produce 37 billion radioactive decays per second (3.7×10^{10} dis/sec), which is the equivalent activity of 1 gram of radium-226. Named after the Polish-born French physicist, Marie Curie (1867–1934) and her French physicist husband, Pierre Curie (1859–1906). A similar unit is the becquerel, defined as 1 disintegration per second and named after another Frenchman, Henri Becquerel.

Daughter products. Elements produced by decay of a radioactive element. Sometimes there are granddaughter and great-granddaughter radionuclides. Not all daughter products are radioactive; some are stable elements.

Depleted uranium. Uranium (U) is a white radioactive metallic element whose atomic number is 92. Depleted uranium usually refers to residue from a nuclear reactor or isotope-separation plant. It is "depleted" because it contains fewer fissile atoms—that is, less U-235—than natural uranium. Natural uranium consists of 99.2745 percent U-238, 0.720 percent U-235, and 0.0055 percent U-234.

Dose. A measure of the quantity or amount of ionizing radiation to which a person has been exposed. Absorbed dose represents the mean energy per unit mass absorbed by matter as a result of exposure to ionizing radiation.

Dosimetry. Measurement of radiation doses with a dosimeter system or instrument. A dosimeter is any device capable of measuring the amount or intensity of either electromagnetic (wave types, such as X or gamma) or particulate (alpha or beta) radiation.

Electromagnetic radiation. A form of radiation that is a wave traveling at the speed of light (3×10^{10} cm/sec = 186,000 mi/sec) in a vacuum and slightly slower in a medium such as water or glass. Examples are X and gamma radiation. In contrast, particle radiation consists of electrons, protons, neutrons, and alpha particles.

Electron. A small particle that can act as a wave (for example, in an electron microscope), one of the stable basic particles of matter. The electron has a negative charge and its mass (at rest) is 9.11×10^{-28} g, equivalent to 0.511 MeV of gamma radiation.

Electron volt (eV). Unit of energy, 0.16 billion-billionth of a joule; the energy gained by an electron as it is accelerated through an electrical potential of 1 volt.

Electrostatic precipitator. An instrument used to collect very small dust particles. As used by the health physicist, it consists basically of an air pump and a source of high voltage (more than 10,000 volts). This high voltage is applied to a sharp-pointed rod (needle) so that it produces an electrical discharge to air at the mouth of a foil-lined cylinder about 2.5 cm in diameter, through which air is drawn. The electrical discharge at the point of the needle ionizes the incoming air so that electrons are attached to dust particles as the air is drawn through the cylinder. The electrically charged dust particles are plated out in the walls of the aluminum foil lining the cylinder. The aluminum foils are removed periodically, and the radioactivity on the attached dust is analyzed.

Epithermal neutron. A neutron with speeds and energies intermediate between those of fast and thermal neutrons. The energy ranges between about 0.02 eV and 100 eV.

Equivalent roentgen (er). Unit of radiation measurement that can be applied to any type of ionizing radiation—alpha, beta, electron, or neutron—and to radiation absorbed or generated in any medium, such as human tissue. It corresponds to the absorbed energy of a roentgen or to 87.7 ergs of energy per gram, irrespective of the medium in which it is produced or absorbed. The roentgen, defined as the quantity of X or gamma radiation such that the associated corpuscular emission per 0.001293 grams of air produces in air ions carrying 1 esu of electricity of either sign, was never a very satisfactory unit. By definition it applied only to X or gamma radiation and that which produced corpuscular (particle) radiation—that is, mostly electrons that were produced in air and absorbed in air to 0°C and standard pressure (760 mm Hg). It did not apply to primary particulate radiation or to radiation absorbed or generated, for example, in human tissue. Eventually the er was replaced by the rad.

Erythema. Redness of skin. The threshold erythema dose was the dose of ionizing radiation for which skin redness just began to make its appearance.

External radiation dose. The dose (expressed in equivalent roentgens, rads, grays, or sieverts) from a source of ionizing radiation located outside the human body.

Fallout. The debris of radioactive dust particles from the detonation of a nuclear weapon that is carried by the wind and settles on the countryside. It comprises fission products such as Sr-90, fissile materials such as Pu-239 or U-235, and induced radionuclides such as Fe-55, Co-60 and Na-24.

Fast neutron. A neutron with energy above about 100 eV and usually with energies of millions of electron volts, MeV.

Fast neutron dose. The density of energy (ergs/gm, joules/kg, roentgens equivalent, rads, rems, grays, or sieverts) in a person or object, such as air, from neutrons.

Fast neutron fission. Nuclear fission produced by a fast neutron. Pu-239 has a large fission cross-section for both thermal and fast neutron fission.

Fissile material. Radionuclides, such as U-235 or Pu-239, that have a large probability or cross-section for fission. For example, the capture cross-section, σ_c, and fission cross-section, σ_f, for thermal neutrons are: $\sigma_f = 586$ b and $\sigma_c = 95$ b for U-235; $\sigma_f = 752$ b and $\sigma_c = 270$ b for Pu-239; and $\sigma_f = 3$ μb and $\sigma_c = 2.7$ b for U-238.

Fission. As used here, the breaking apart of a fissile atom into fission products and neutrons. For example, during the fission of the U-235 nucleus by a neutron we might have the following reaction:

$$^{1}_{0}n + ^{235}_{92}U \xrightarrow{f} ^{143}_{56}Ba + ^{91}_{36}Kr + 2^{1}n + E$$

Note that in all cases the sum of the atomic masses (superscript numbers) on one side of the equation must equal the sum of atomic masses on the other side, and the sum of the atomic numbers (subscript numbers) on one side must equal the sum of atomic numbers on the other. Here E is the energy of the explosion, which is about 200 MeV. Ba is the fission product barium with a half-life of 14.3 seconds, and Kr is the fission product krypton with a half-life of 8.6 seconds. The equation can be written in other ways, for example, n + U-235 → Ba-143 + Kr-91 + 2 n.

Flux. A flow of particles, such as electrons, neutrons, or protons, or of photons (waves), such as X-rays or gamma rays.

Gamma radiation. A high-energy, wave-type radiation produced in nuclear reactions. It is similar to X-rays except that it originates from a nuclear transition while X-rays are produced by the deceleration of electrons when they strike a heavy target. Gamma radiation originates as monochromatic (single wavelength) radiation, while X-rays consist of a spectrum of wavelengths.

Gaseous diffusion. One of the methods by which an isotope of one atomic mass, such as U-235, can be sorted out from a mixture, such as the natural mixture of U-238, U-235, and U-234, and thus concentrated. This was the principal method used at the large K-25 plant and other AEC plants to concentrate U-235 for weapons. In gaseous diffusion a gaseous compound of uranium is passed through a long chain (many miles) of barriers, which by diffusion and millions of collisions of gas molecules results in the separation of the isotopes.

Geiger-Müller counter (Geiger counter, GM counter). A container (usually a glass tube of various sizes and shapes containing a metal cylinder and a central wire across which a high voltage is applied) that gives a clicking sound and a sharp electrical pulse (discharge) when struck by an ionizing particle, such as an electron or a photon—gamma ray—of sufficient energy to produce an electrical discharge. The tube contains a low-pressure gas mixture that immediately quenches the electrical discharge to prevent a continuous DC current flow. It is one of the simplest and most reliable instruments used by the health physicist. Care must be taken in its use because in an intense field of ionization (e.g., 10,000 rad/hr), the clicks may come so fast that the counter is paralyzed—that is, it ceases to click and the output meter reads zero.

Graphite reactor. A nuclear reactor containing blocks of graphite (a soft, black, lustrous form of carbon) interspaced with slugs of uranium. The graphite is used to moderate the speed of the neutrons produced during fission. It is used because its principal isotope, C-12 at 98.892 percent, has a low capture cross-section, $\sigma_c = 0.0035$ b, and its other isotope, C-13 at 1.108 percent, a capture cross-section of $\sigma_c = 0.0014$ b. To be used in a reactor it must be extremely pure carbon. For example, Co-59 (100 percent) has $\sigma_b = 20.7$ b. The slugs of uranium are canned in aluminum tubes. *See also*: Air-cooled, graphite-moderated reactor (AGR) and water-cooled, graphite-moderated reactor (WGR).

Gray unit (Gy). The amount of dose that corresponds to 100 rads or to 10,000 ergs per gram of the exposed material. Named in honor of British scientist Hal Gray. *See also*: Bragg Gray effect.

Half-life. Amount of time necessary for a radioactive substance to lose 50 percent of its activity through the decay process.

Heavy water. Deuterium oxide, or water in which the hydrogen atoms are replaced by the heavier isotope deuterium, ^{2}H. It is well suited as a moderator to slow down neutrons in nuclear reactors because $\sigma_c = 0.0005$ b for H-2 and 0.00019 b for O-16.

Hot particle problem (HPP). Nuclear reactors such as those at Oak Ridge and Hanford released small dustlike particles, such as iron oxide or graphite, that were radioactive and contained fission products, such as I-131, Cs-135, Sr-90, uranium, and plutonium (Pu-239, Pu-240). Since the dose rate of a radioactive source decreases inversely with the distance from the source, the dose rate close to a small hot particle can be extremely large, especially if the particle contains an alpha emitter like Pu-239.

Hurst neutron dosimeter. A portable fast neutron dosimeter developed by G. S. Hurst of the ORNL Health Physics Division. By using a proportional counter it could measure the fast neutron dose in a field of mixed radiation. It was important to have such an instrument for two reasons: first, because the permissible dose of fast neutrons is much less than that for other types of external exposure, and second, because at times it was critical to know whether the exposure was from fast neutrons or other types of radiation. The presence of fast neutrons was clear evidence of a nuclear reaction, and this could mean that a criticality accident was under way.

Integrated total body dose. The energy in ergs per gram to the total body from exposure to ionizing radiation, whether from external or internal dose or a combination of both. When expressed in rads, it is the sum of the ergs delivered to the total body divided by the mass of the total body in grams and divided by 100.

Internal dose. The dose from radionuclides within the human body. It may come from radionuclides in the gastrointestinal tract and respiratory system or from radionuclides deposited in body organs, for example, I-131 in the thyroid.

Internal radiation. Radiation produced by radionuclides located inside the human body.

Ion chamber. An apparatus used to measure the amount of ionization produced by radioactivity in air.

Ionizing radiation. Photons (X-rays and gamma rays) and ionizing particles (alpha, beta, protons, heavy ions) with sufficient energy to produce ionization directly as they pass through air. Ionization is the freeing of electrons from an atom. In the process ion pairs are formed—the electron with a negative charge and the positive ion. Ion pairs are produced in air when the energy of the radiation is about 15 eV. Ultraviolet radiation is in the intermediate energy region (3.18–124 eV) between ionizing and nonionizing radiation. Visible light (red 1.57 eV to violet 3.18 eV) is classed as nonionizing radiation.

Isotopes. Similar to radionuclides, that is, nuclides or isotopes that are radioactive. Isotopes that are radioactive and that do not occur in nature are usually referred to as radionuclides rather than radioisotopes. The number of protons in the nucleus of an atom defines the element. In nature the number of neutrons in the nucleus is about equal to the number of protons, but it can vary considerably and can be made to vary still more by applying high energy to the nucleus of an atom. For example, both C-12 and C-13 occur in nature, but neutrons can be subtracted to form C-11, C-10, or C-9, or added to form C-14, C-15, C-16, and so on. Adding neutrons to or subtracting them from the nucleus always forms radionuclides. C-14 also occurs in nature and is produced by cosmic radiation and by naturally occurring radioisotopes, such as Ra-226, U-238, Th-232, etc. *See also*: Radionuclide.

Lauritsen electroscope. One of the earliest and most reliable radiation monitoring and portable survey instruments used by health physicists at Chicago and Oak Ridge, developed by C.C. Lauritsen. It was a small, boxy instrument containing an electroscope and batteries. The electroscope was a thin-walled aluminum ionization chamber connected electrically to a quartz fiber suspension system that was charged by the batteries. The motion of the fiber could be determined by viewing it through an incorporated microscope and noting its motion with reference to a scale in the background marked off in milliroentgens. It was used to measure the radiation dose of X, gamma, and beta radiation. The dose rate was measured by timing the fiber with a stopwatch as it moved across the scale. It gave rather accurate values of rads but was cumbersome to use. It contained a tiny wooden window that could be opened to measure the sum of beta plus gamma dose. The closed-window reading was subtracted to determine the gamma or X-ray dose.

Linear hypothesis. As used in this book, one of three hypotheses regarding the development of cancer following exposure to ionizing radiation: the linear hypothesis, the threshold hypothesis, and the supralinear hypothesis. The linear hypothesis assumes that the probability of developing a radiation-induced cancer increases linearly with the dose. That is, doubling the dose doubles the cancer risk, tripling the dose triples the risk, and so forth. The threshold hypothesis assumes that there is no increase in cancer incidence in a population or in the risk to a person unless the dose exceeds a poorly defined threshold dose. This threshold is usually considered to be one or two times the natural background dose or 100–200 mrem. The supralinear hypothesis assumes that the probability of developing a radiation-induced cancer increases more rapidly with the dose at low doses than at high doses.

Liquid metal fast breeder reactor (LMFBR). A nuclear reactor that during its operation produces fissile material (U-235, Pu-239, U-233). It may breed Pu-239 from U-238 or U-233 from Th-232 by the following reactions:

$$^{238}_{92}U + ^{1}_{0}n \rightarrow ^{239}_{92}U\ (23.5m) \qquad ^{232}_{90}Th + ^{1}_{0}n \rightarrow ^{233}_{90}Th\ (22.3m)$$

$$^{239}_{92}U \rightarrow ^{239}_{93}Np\ (2.355d) + ^{0}_{-1}\beta \qquad ^{233}_{90}Th \rightarrow ^{233}_{91}Pa\ (27.0d) + ^{0}_{-1}\beta$$

$$^{239}_{93}Np \rightarrow ^{239}_{94}Pu\ (24{,}110y) + ^{0}_{-1}\beta \qquad ^{233}_{91}Pa \rightarrow ^{233}_{92}U\ (1.59 \times 10^{5}y) + ^{0}_{-1}\beta$$

Maximum permissible concentration (MPC). The concentration in air, water, and food that will deliver the maximum permissible dose to the total body or to an individual organ of the body.

Maximum permissible exposure (MPE). A limit on the dose of ionizing radiation (measured in rems) that is permitted in a given time interval and is considered not to cause appreciable or unacceptable harm. It may be applied to external or internal dose or both. Values of MPE at the international level are set by the ICRP and in the United States by the NCRP. Other agencies, such as the Nuclear Regulatory Commission or the Federal Radiation Council, set their own levels of MPE, which in principle are equal to or less than those set by the ICRP or the NCRP. The 1997 MPE levels set by ICRP are 2 rem/y (20 mSv/y) for occupational exposure averaged over five years and 100 mrem/y (1 mSv/y) for members of the public. *See also*: Maximum permissible concentration (MPC).

Meson. A particle of intermediate mass between that of the electron and proton or neutron. The pi meson is one of the particles observed in nuclear reactions. The positive or negative meson has a mass 273 times that of the electron and a charge the same as that of an electron. There also are neutral mesons and mu (μ) mesons. Mesons require extremely high energies for their production. They make up the major part of secondary cosmic radiation.

Minometer. A type of photometer used to measure the darkening of dental films used in film badges. The darkening was due to exposure to ionizing radiation.

Mixed radiation. Radiation of different types. In many locations at ORNL there was the potential of simultaneous exposure from mixed radiation. Health physics surveyors needed to know not only the dose rate, rad/h, at a given location but also the dose rate from each type of radiation. Most of the survey meters were designed to measure the absorbed (physical) dose rate, and since the neutron dose rate in rad/h (or Gy/hr) had to be multiplied by 30 to convert to rem/h (or Sv/h), it was not sufficient to know only the dose rate of mixed radiation in rad/h or Gy/h. Also the dose from beta radiation was only to the surface of the body—the hands and face—so it was important to know how much of the dose rate was due to beta and how much to gamma radiation.

Moderator. Any type of substance used to slow down neutrons in a nuclear reactor, thereby making it more probable to cause fission of uranium. Light elements, such as graphite, beryllium, and deuterium, can be used as moderators. For efficient moderation one chooses a substance with a low atomic number and low capture cross-section.

Molten salt thermal breeder (MSTB). One of the many types of reactors developed by ORNL. The MSTB is operated at high temperature for the production of power, using fissile material (U-235, U-233, Pu-239) and fertile material (material such as U-238 or Th-232 that can be converted to fissile material by neutron exposure) in the form of a fluoride salt dissolved in the coolant. The coolant is a molten salt mixture, such as beryllium fluoride and lithium fluoride. Unlike the MSTB, the LMFBR did not contain the fused salts (with low atomic mass) that would slow down or thermalize the neutrons, which during fission are born with high (fast) energy. *See also*: Liquid metal fast breeder reactor (LMFBR).

Natural background radiation. Radiation in the environment, consisting of cosmic radiation that increases with elevation and includes naturally occurring radiation from radionuclides, such as U-238, Th-232, U-235, and their daughter products, and from a few other naturally occurring radionuclides.

Neutron. A fundamental particle that has about the same mass as the proton and a zero charge.

Nucleons. Neutrons and protons, which serve as the basic building blocks of atomic nuclei.

Nucleus. The core of an atom, containing one or more protons and a number of neutrons. The number of neutrons may be zero only in the case of hydrogen; for other elements, the number of neutrons may be equal to, less than, or more than the number of protons. The number is always greater for elements of large atomic mass.

Particulates. Fine solid particles, individually dispersed in gases and emitted in exhaust stacks of nuclear reactors.

Pencil meters. Small pencillike meters, consisting of a hollow cylinder containing a central electrode that is electrically insulated from the wall of the cylinder. Before use an electrical charge is placed between the central electrode and the cylinder wall, both of which must be made of a conductive material, such as metal or graphite. Pencil meters are customarily worn on clothing. When exposed to ionizing radiation, the amount of electrical discharge is proportional to the dose, which can be read with an electrometer calibrated in rads. Also called pocket condenser meter.

Photographic film badge. A small badge that can be worn on a person's clothing, containing dental films. The film emulsion is darkened when exposed to ionizing radiation. After they have been worn for a given period of time, film badges are unwrapped and developed in a photographic darkroom, then read with a photometer that measures the amount of light that will pass through the film. The photometer is calibrated in millirad. The larger the exposure, the higher the mrad reading on the photometer.

Pile. The early name given to a nuclear reactor. It was used because the early reactors at Chicago and Oak Ridge were actually piles of graphite blocks interspaced with slugs of natural uranium.

Plutonium (Pu). The element of atomic number 94. It can be produced from uranium in reactors and by high-voltage acceleration. Until Pu-238 was first produced by Joseph Hamilton in the Berkeley cyclotron in 1932, plutonium existed in the natural environment only in minute amounts and had not previously been detected. Many radionuclides of plutonium have been produced, the most notable of which are Pu-239 (24,110 y), Pu-240 (6,537 y), Pu-238 (87.74 y) and Pu-241 (14.4 y). In insoluble form, it is a lung hazard to human beings; in soluble form, it concentrates in the periosteal and endosteal tissue and is likely to produce bone cancer.

Pocket fiber electrometer. A pen-sized pocket dosimeter, a small electrometer about the size and shape of the pencil meter. *See also*: Pencil meter.

Portable fast neutron survey instruments. *See*: Hurst neutron dosimeter.

Pressurized water reactor (PWR). A reactor in which the water is at high temperature and at high pressure; the principal type of power reactor in use in the world today to produce electricity and propel nuclear submarines. The other type of reactor used in considerable numbers in the United States is the boiling water reactor (BWR).

Prompt critical. A reaction that occurs instantaneously without the aid of the delayed neutrons that keep the reactor operating at steady power. The delayed neutrons are emitted spontaneously from a fissile nucleus as a consequence of excitation left from preceding radioactive decay events.

Proportional counter. A counter similar in appearance to a Geiger counter but operated at a voltage that produces ionization by collision but at low enough energy to exclude counts from X and gamma rays. It has a gas mixture (argon and methane) and pressure differing from that used in a Geiger counter. It can be designed to count alpha particle radiation or fast neutrons in the presence of mixed radiation.

Proton. One of the fundamental atomic particles, having a positive charge and a mass of 1.6726×10^{-24} g. It makes up most of the primary cosmic radiation. The neutron when outside the nucleus of an atom decays with a half-life of 10.3 minutes into a proton and a beta particle.

Quality factor (Q). A term used in dosimetry to convert absorbed energy units (reps, grays) into units of harm (rems, sieverts). In the case of alpha particles or fast neutrons, the ICRP recommends using a quality factor of 30. So for alpha or fast neutron dose we have 1 rad = 30 rem or 1 gray = 30 sieverts.

Rad. The basic unit of absorbed dose (radiation absorbed dose), referring to absorbed energy in tissue (100 ergs/gm or .01 J/kg).

Rad equivalent man (rem). Unit of radiation dose equivalent used for safety purposes. The dose equivalent is equal to absorbed dose in rad times the quality factor and any other modifying factors. The rem has been replaced by the sievert (Sv) unit.

Radiation syndrome. The series of human body changes and events following exposure to a high dose of ionizing radiation in a single event or over a short time. Some of the events are prostration, fatigue, nausea, sweating, fever, infection, hemorrhaging, diarrhea, cessation of intestinal movements, cardiovascular collapse, lethargy, loss of consciousness and perception, severe pain, and death at the higher doses. The mid-lethal dose is commonly taken as 400 rad (4 gray), but Joseph Rotblat in a reanalysis of the Japanese atomic bomb survivor data found a mid-lethal dose of 154 rads.

Radioisotopes. *See*: Isotopes.

Radionuclides. *See*: Isotopes.

Roentgen (R). An obsolete unit of exposure of ionizing radiation. *See also*: Equivalent roentgen.

Roentgen equivalent physical (rep). Same as equivalent roentgen; a unit corresponding to the same energy deposition as that of a roentgen.

Sievert unit (Sv). Unit of potential harm from exposure to ionizing radiation. A radiation dose adjusted to living tissues.

1 Sv = 1 Gy × Q where Q = quality factor

1 Sv = 100 rem

Somatic cell. Any cell of the body except a germ cell. The sperm or oocyte (an egg before maturation or becoming mature) cells are germ cells.

Spectrometer. An instrument for identifying and measuring intensities of various wavelengths or frequencies of radiation.

Spontaneous fission. The fission of the atom when no high-energy particles (neutrons, protons) or photons enter the nucleus.

Stable element. Isotope of an element that is not radioactive.

Subcritical reactor (assembly). A reactor in which the neutron multiplication constant is less than one, so that a self-sustaining chain reaction cannot be maintained. This type of reactor is commonly used in universities for training nuclear engineers.

Supralinear hypothesis. *See*: Linear hypothesis.

Teratogenic changes. Usually refers to abnormal development during the first two trimesters of pregnancy. These changes are distinguished from genetic damage that results from abnormal changes starting in the gonads rather than the fetal tissue.

Terrestrial radiation. Radiation from naturally occurring radionuclides, such as U-238 and Th-232 and their daughter products and K-40. Distinguished from cosmic radiation. *See also*: Natural background radiation.

Thermal neutron. A neutron whose energy is identical to that of a normal environment (room temperature). A thermal neutron or slow neutron has the same kinetic energy as the air molecules at room temperature. Its energy thus is about 0.025 eV.

Thermonuclear reaction. A process in which very high energy brings about the fusion of light nuclei with accompanying liberation of very high energy.

Threshold dose. The minimum dose that produces a given effect, such as slight reddening of skin from X-ray exposure of about 20 R to persons with very sensitive skin.

Threshold hypothesis. *See*: Linear hypothesis.

Uranium slugs. Small cylinders of natural uranium contained in small aluminum cans welded shut at the ends, which were inserted in the graphite reactor at Oak Ridge National Laboratory.

Water-cooled, graphite-moderated reactor (WGR). Type of nuclear reactor cooled by fast-flowing water and moderated by graphite. At the Hanford weapons facility, the natural uranium was contained in slugs positioned in channels in a graphite matrix. Water flowed over the slugs to provide cooling. The speed of the neutrons produced by the fission of U-235 was slowed down (moderated) by the graphite to increase the probability that the neutrons would be captured by the uranium to produce Pu-239 by the following reactions:

$${}^{238}_{92}U + {}^{1}_{0}n \rightarrow {}^{239}_{92}U\ (23.5m) \rightarrow {}^{239}_{93}Np\ (2.355\ d) + {}^{0}_{-1}\beta$$

$${}^{239}_{93}Np \rightarrow {}^{239}_{94}Pu\ (24{,}110y) + {}^{0}_{-1}\beta$$

See also: Air-cooled, graphite-moderated reactor (AGR).

Weapons-grade uranium. High-purity U-235. Natural uranium is 99.275 percent U-238, 0.720 percent U-235, and 0.0055 percent U-234. Only U-235 sustains the fission process with emission of 200 MeV of energy during fission, but it cannot be used as a weapon unless a critical mass of U-235 can be brought together very rapidly. The fission stops instantly when the U-235 components are blown apart, because it is the neutrons that sustain the fission, and the tighter the U-235 assembly, the fewer neutrons escape without contributing to more fissions. For weapons-grade uranium one wishes to remove as much of the U-238 and U-234 from the U-235 as is cost-effective; good weapons-grade uranium might be as much as 99 percent U-235. The half-life of U-235 is 0.71 billion years.

Wigner energy. Energy stored in a crystal or crystal-like substance, such as diamond or pure graphite; named after its discoverer, Eugene Wigner. In the presence of ionizing radiation, electrons and electron holes can be raised to energy levels above the ground state. Then in the presence of heat they are released, dropping back to the ground state and increasing the temperature. At low temperature increases they are released from the lower energy levels, and as the temperature is increased, they are released from higher and higher energy levels. In nuclear reactors containing graphite as a neutron energy moderator, large amounts of energy can be accumulated and stored in the graphite waiting to be released suddenly if there is a large increase in temperature. This can lead to melting of fuel elements, blockage of the reactor cooling system, and fires in the graphite. The situation in graphite-moderated reactors is ameliorated by scheduled slow heating of the graphite and proper dissipation of the Wigner energy. This stored energy in the graphite (Wigner effect) is the bane of the nuclear industry, causing the Windscale accident and undoubtedly contributing to other nuclear accidents.

X-ray radiation. A form of electromagnetic radiation, having two types, characteristic and continuous. Characteristic X-rays have a specific energy as a result of rearrangement of electrons in the inner shells of atoms of high atomic number. They are seen as line spectra. Continuous X-rays are produced by the gradual slowing down of an X-ray beam of electrons when they strike the heavy metal of the X-ray target in the X-ray tube; they appear as a continuous spectrum of energies (wave lengths) beginning with the sharp maximum down to the lowest X-ray energy. Many radionuclides give off X-rays as well as other forms of radiation.

BIBLIOGRAPHY

Allen v. United States. 588 F. Supp. 247 (D. Utah 1984).

Allen v. United States. 816 F.2d 1417 (10th Cir. 1987).

Allen v. United States. Transcript of trial proceedings.

Cochran, Thomas B., and Robert S. Norris. "A First Look at the Soviet Bomb Complex." *Bulletin of the Atomic Scientists* (May 1991): 25–31.

Donovan, Robert J. *Conflict and Crisis: The Presidency of Harry S. Truman, 1945–1948*. Vol. 1. New York: W. W. Norton, 1977.

Gallagher, Carole. *American Ground Zero: The Secret Nuclear War*. Foreword by Keith Schneider. New York: Random House, 1993.

Goldberg, Stanley. "Groves Takes the Reins." *Bulletin of the Atomic Scientists* (December 1992): 32–40.

Griffiths, Joel, and Richard Ballantine. *Silent Slaughter*. Chicago: Henry Regnery, 1972.

Hawkes, Nigel, Geoffrey Lean, and David Leigh et al. *Chernobyl: The End of the Nuclear Dream*. New York: Vintage Books, 1987.

Health Physics Research at Oak Ridge National Laboratory. Prepared by Oak Ridge National Laboratory for U.S. Atomic Energy Commission, n.d. (c. 1967).

"Interim Report of the Advisory Committee on Human Radiation Experiments." October 21, 1994. Washington, D.C.: U.S. Government Printing Office.

Larsen, R. P., and R. D. Oldham. "Plutonium in Drinking Water: Effects of Chlorination on Its Maximum Permissible Concentration." *Science* 201 (September 15, 1978): 1008.

Litton, Gary W. "What Has America Done" (March 1994). Unpublished paper in possession of the authors.

Lowe, Alexandra Dylan. "The Price of Speaking Out." *American Bar Association Journal* (September 1996): 48–53.

Mancuso, Thomas F., Alice Stewart, and George Kneale. "Radiation Exposures of Hanford Workers Dying from Cancer and Other Causes." *Health Physics* 33 (November 1977): 369–85.

Manhattan Engineer District Records. Harrison-Bundy Files.

"Maximum Permissible Amounts of Radioisotopes in the Human Body and Maximum Permissible Concentrations in Air and Water." National Bureau of Standards Handbook 52. Washington, D.C.: U.S. Dept. of Commerce, 1953.

"Maximum Permissible Body Burdens and Maximum Permissible Concentrations of Radionuclides in Air and in Water for Occupational Exposure." NCRP Report 22, National Bureau of Standards Handbook 69. Washington, D.C.: U.S. Dept. of Commerce, 1959.

Medvedev, Grigori. *The Truth about Chernobyl.* Translated by Evelyn Rossiter. New York: Basic Books, 1989.

Miscellaneous Historical Document Collections 345 and 671. Harry S. Truman Library, Independence, Missouri.

Modan, Baruch, Hannah Mart, Dikla Baidatz, Ruth Steinitz, and Sheldon G. Levin. "Radiation-Induced Head and Neck Tumors." *Lancet* (February 23, 1974): 277–79.

Monitored Retrievable Storage Review Commission. "Nuclear Waste: Is There a Need for Federal Interim Storage?" (November 1, 1989). Washington, D.C.: Monitored Retrievable Storage Commission.

Moore, Mike. "The Incident at Stagg Field." *Bulletin of the Atomic Scientists* (December 1992).

Morgan, Karl Z. "The Body Burden of Long-Lived Isotopes." *Archives of Environmental Health* 8, no. 1 (January 1964): 86.

———. "Changes in International Radiation Protection Standards." *American Journal of Industrial Medicine* 25 (1994): 301–7.

———. "Determination of Exposures." In *Radiation Hygiene Handbook*, edited by Hanson Blatz. New York: McGraw-Hill, 1959. Pp. 14-1–14-19.

———. "Do Low-Level Radiation Health Data Justify Fear or Contribute to Phobia?" *Physics Today* (August 1992): 9.

———. "Health Physics." In *American Institute of Physics Handbook.* New York: McGraw-Hill, 1957; 2nd ed., 1963. Pp. 8-312–8-327.

———. "Health Physics." In *Encyclopedia of Physics*, edited by R. M. Besancon. New York: Reinhold Publishing, 1966. Pp. 306–8. 2nd ed., 1974. Pp. 404–6. 3rd ed., 1985. Pp. 542–46.

———. "Health Physics." In *Nuclear Engineering Handbook.* New York: McGraw-Hill, 1958. Pp. 7-22–7-60.

———. "Health Physics: Its Development, Successes, Failures and Eccentricities." *American Journal of Industrial Medicine* 22 (1992): 125–33.

———. "Human Radiation Studies: Remembering the Early Years. Oral History of Health Physicist Karl Z. Morgan, Ph.D." U.S. Department of Energy, Office of Human Radiation Experiments, June 1995.

———. "ICRP Risk Estimates—An Alternative View." In *Radiation and Health: The Biological Effects of Low-level Exposure to Ionizing Radiation.* Chinchester: John Wiley and Sons, 1987. Pp. 125–54.

———. "The International Commission on Radiological Protection Made a Bad Mistake When It Increased MPC Values in 1979–88 and in Its Draft Report of October 15, 1990 (Now ICRP-61). This Mistake Is Only Partially Corrected." *International Perspectives in Public Health* 7 (1991): 15–19.

———. "Ionizing Radiation Exposure." In *Environmental Problems in Medicine.* Springfield, Ill.: Charles C. Thomas, 1974.

———. "A Juggling Act with Values of Maximum Permissible Concentration of Radionuclides in Air and Water." *Health Physics* 62, no. 3 (1992): 264–66.

———. Letter to *The Oak Ridger.* December 19, 1993.

———. "Permissible Quarterly Intakes of Radionuclides." In *Handbook of Chemistry and Physics*, edited by Robert C. Weast. 52nd ed. Cleveland: Chemical Rubber Co., 1971–72. P. B-542.

———. "Present Status of Recommendations of the International Commission on Radiological Protection, National Council on Radiation Protection and Federal Radiation Council." In *Health Physics*, vol. 2, part 1. Oxford: Pergamon Press, 1969. Pp. 11–34.

———. "Radiation Protection." In *Nuclear Reactors for Industry and Universities*. Pittsburgh: Instruments Publishing Co., 1954. Pp. 38–50.

———. "Radiation: Protection and Health Physics." In *Medical Physics*, edited by Otto Glasser. Chicago: Year Book Publishers, 1950. Vol. 2, pp. 766–74.

———. "Suggested Reduction of Permissible Exposure to Plutonium and Other Transuranium Elements." *American Industrial Hygiene Association* 36, no. 8 (August 1975): 567.

———. "Underestimating the Risks." In *Nuclear Power: Both Sides*. New York: W. W. Norton, 1982. Pp. 35–46.

Morgan, Karl Z., and W. M. Nielsen. "Shower Production under Thick Layers of Various Materials." *Physical Review* 52, no. 6 (September 15, 1937): 564.

Morgan, Karl Z., W. S. Snyder, and M. R. Ford. "Relative Hazard of Various Radioactive Materials." *Health Physics* 10, no. 3 (1964): 151.

Morgan, Karl Z., I. H. Tipton, and M. J. Cook. "A Summary of the Data That Was Used in the Revision of the Internal Dose Recommendations of the International Commission on Radiological Protection." In *Health Physics*. New York: Pergamon Press, 1959. Vol. 1, pp. 17–28.

Morgan, Karl Z., and James E. Turner. "Health Physics." In *American Institute of Physics Handbook*, 3rd ed. New York: McGraw-Hill, 1972. Pp. 8-291–8-315.

Morgan, Karl Z., and James E. Turner, eds. *Principles of Radiation Protection*. New York: John Wiley and Sons, 1967.

———. *Principles of Radiation Protection*. New York: Robert E. Krieger, 1973.

Nielsen, W. M., J. E. Morgan, and Karl Z. Morgan. "The Rossi Transition Curve for Small Angle Showers." *Physical Review* 55, no. 11 (June 1, 1939): 995.

Nielsen, W. M., and K. Z. Morgan. "The Absorption of the Penetrating Component of the Cosmic Radiation." *Physical Review* 54, no. 4 (August 15, 1938): 245.

Nielsen, W. M., C. M. Ryerson, L. W. Nordheim, and Karl Z. Morgan. "Differential Measurement of the Meson Lifetime." *Physical Review* 59, no. 7 (April 1, 1941): 547.

———. "A Measurement of Mesotron Lifetime." *Physical Review* 57, no. 2 (January 15, 1940): 158.

"Recommendations of the International Commission on Radiological Protection" (as amended 1959 and revised 1962). ICRP Pub. 6. New York: Macmillan, 1964.

"Report of Committee II on Permissible Dose for Internal Radiation: Recommendations of the International Commission on Radiological Protection." ICRP Pub. 2. London: Pergamon Press, 1959.

Silkwood v. Kerr-McGee Corporation. 485 F. Supp. 566 (W.D. Okla. 1979).

Silkwood v. Kerr-McGee Corporation. Transcript of trial proceedings.

Spence, Gerry, and Anthony Polk. *Gerry Spence Gunning for Justice.* Garden City, N.Y.: Doubleday, 1982.

Stewart, Alice, Josefine Webb, Dawn Giles, and David Hewitt. "Malignant Disease in Childhood and Diagnostic Irradiation in Utero." *Lancet* (September 1, 1956): 447.

Stillwagon, G. B., and Karl Z. Morgan. "In-Situ Dosimetry of Plutonium-239 in Bone Using Polycarbonate Foils and Electrochemical Etching." *International Perspective in Public Health* 5, no. 1 (summer 1989): 7.

Tamplin, Arthur R., and Thomas B. Cochran. "The Hot Particle Issue: A Critique of WASH 1320 as It Relates to the Hot Particle Hypothesis." Natural Resources Defense Council, November 1974. Pp. 1–47.

———. "A Report on the Inadequacy of Existing Radiation Protection Standards Related to Internal Exposure of Man to Insoluble Particles of Plutonium and Other Alpha-Emitting Hot Particles." Natural Resources Defense Council, February 1974. Pp. 1–52.

U.S. Senate, Committee on Governmental Affairs. "Human Radiation and Other Scientific Experiments: The Federal Government's Role." 103d Cong., 2d sess., January 25, 1994.

Weinberg, Alvin M. *The First Nuclear Era.* New York: American Institute of Physics Press, 1994.

White House Confidential Files. Harry S. Truman Library, Independence, Missouri.

Index